# 100
# MILLION YEARS
# OF FOOD

# 100 MILLION YEARS OF FOOD

## WHAT OUR ANCESTORS ATE AND WHY IT MATTERS TODAY

## STEPHEN LE

PICADOR

New York

picadorusa.com
twitter.com/picadorusa • facebook.com/picadorusa
picadorbookroom.tumblr.com

Picador® is a U.S. registered trademark and is used by St. Martin's Press under license from Pan Books Limited.

For book club information, please visit
facebook.com/picadorbookclub or e-mail marketing@picadorusa.com.

Designed by Steven Seighman

Library of Congress Cataloging-in-Publication Data

Le, Stephen.
One hundred million years of food : what our ancestors ate and why it matters today / Stephen Le. — First edition.
    p. cm
  ISBN 978-1-250-05041-0 (hardcover)
  ISBN 978-1-250-05042-7 (e-book)
 1. Natural foods.  2. Food—History.  3. Nutrition.  4. Prehistoric peoples—Food.  I. Title.  II. Title: 100 million years of food.
  TX369.L396 2015
  641.3—dc23

2015029501

Our books may be purchased in bulk for promotional, educational, or business use. Please contact your local bookseller or the Macmillan Corporate and Premium Sales Department at 1-800-221-7945, extension 5442, or by e-mail at MacmillanSpecialMarkets@macmillan.com.

First Edition: February 2016

10  9  8  7  6  5  4  3  2  1

*for my father and in memory of my mother*

# CONTENTS

# INTRODUCTION

## *What Should We Eat and How Should We Live?*

Around the world, people are increasingly beset with vexing conditions like obesity, type 2 diabetes, gout, hypertension, breast cancer, food allergies, acne, and myopia. Dubbed the "diseases of Western civilization," these conditions have become more common in recent centuries and decades, with emigrants to more affluent regions of the world often at higher risk for developing afflictions as they adopt the customs and diet of their new home. It's true that better medical care and longer life spans are exposing people to new kinds of diseases. Still, the rapid advance in these conditions signals that something else is going wrong—but what?

The principal argument of *100 Million Years of Food* is that a plethora of health woes have surfaced in contemporary times due to our alterations to our ancestors' diets, lifestyles, and environments. I will explain how our ancestors used to eat and live and will offer practical suggestions for tweaking ancestral habits and inserting them into our daily lives to avoid or delay the onset of major chronic diseases.

Eating and living like our ancestors seems common sense, and many books have attempted to relate how this can be done. Unfortunately, there is great disagreement about what aspects of ancestral eating and living should be adopted. For example, in some versions of ancestral diets, readers are advised to avoid agricultural staples like bread, rice, beans, and milk in favor of meat and vegetables, while others argue that traditional American farm diets that included bread, beans, and milk are best for health.

My parents immigrated from Vietnam to Canada in the 1960s. They met at a college in Montreal and brought up three boys in the maple-lined suburbs of Ottawa. Due to my obsession with insects and other critters, I was designated the family science nerd. I received a micro-scope as my Christmas present in fourth grade, which I used to pe-ruse squished mosquitoes on glass slides. The following year, a paperback edition of Charles Darwin's *On the Origin of Species* arrived under the Christmas tree. The letters in the book were extremely small, and the language was unintelligible to an elementary school kid, but I sensed there was something revolutionary between the pages.

When I was eight years old, my grandmother arrived from Viet-nam. When I first met her she was wrapped in a shawl, hunched and wizened, a mage. Grandma opened her luggage and handed me a belt wrought from links of varnished wood, along with an exquisite slice of dried banana, the honeylike taste of which lingers with me to this day. During a school holiday when my younger brother and I visited the apartment that Grandma shared with my aunt, I found no high-energy snacks of nuts and dried fruits, no fruit bars, no yogurt, no blaring TV, just a rocking chair, a few shelves of incomprehensible books, a bottle of fish sauce, a rice cooker, some crumpled old linen towels, and the soft light of day streaming through the window. Oblivious to official pronouncements on nutrition—she spoke no English—Grandma went on living the way she had always lived, eating the way her ancestors had: a bowl of rice, clumps of dried shredded pork, a sprinkling of fish sauce, and a pile of stir-fried spin-ach. I was glad to head back home that evening to my video games and high-energy snacks.

The significance of my grandmother's diet didn't hit me until I visited Vietnam for the first time when I was twenty-five. Wherever I traveled in the country, I drew attention because I towered over the local Vietnamese. Despite being under average height and weight among my peers in Canada at 5'8" and 150 pounds, I was still four inches taller and twenty-seven pounds heavier than my peers in northern Vietnam, who were in turn one inch taller and eighteen pounds heavier than their peers a generation earlier.[1] Why the rapid change? When I was invited to meals around Vietnam, hosts often prepared a feast with pork, chicken, or fish in bountiful proportions. However, when I showed up unexpectedly at mealtimes with ordinary folks, I glimpsed the real side of everyday rural cuisine: rice, paddy or jungle vegetables, fermented fish or soybean sauce for flavoring, tofu, and bits of crab and fish. North American diets rich in meat and milk had evidently made me taller, but as I later learned, such fare also put people who consumed it at greater risk of certain chronic diseases.

Back in Canada, my mother developed breast cancer, which metastasized to her lungs. After I finished my doctoral studies in Los Angeles, I spent three months at home, helping to tend to my mother's needs. She passed away at the age of sixty-six, only two years after her own mother passed away at ninety-two. In the aftermath, everyone in our family had a different means of coping. Dad channeled his energies into community activities; my brothers had their wives and children to occupy them. I decided to focus on researching ancestral diets and lifestyles and learning about the risk factors behind breast cancer and other diseases commonly associated with Western civilization.

I studied biological anthropology (the study of human evolution) at the University of California, Los Angeles (UCLA), then spent two years researching food and food-related illnesses around the world, observing and sampling what people ate and chatting with food producers, health experts, and other people connected to eating and nutrition. I began to understand why health experts disagree sharply on which foods are healthy. First, there is a lot of confusion over the difference between short-term and long-term health. A diet that makes a person taller, a weight lifter stronger, and a woman more fertile is healthy to a degree, but generally not a diet that would make a person live longer. This makes perfect sense from an evolutionary point of view, as will be discussed in this book, but the distinction between short-term and long-term health is often overlooked or misinterpreted by many health researchers. This brings us to the second major mental stumbling block encountered by nutritionists and food writers: the tendency to ignore or misunderstand human evolution and to focus instead on simplistic models of human nutrition and physiology. Trying to understand human nutrition and health without understanding evolution is like trying to eavesdrop on a snippet of conversation without knowing the context—it makes little sense or can be very misleading.

As a result of my observations and reinterpretation of scientific studies related to food, I discovered a series of measures that people can take to improve their health. These interventions are discussed throughout the book and summarized in the afterword. However, there are three steps that I consider the most important to improve the health of someone living in modern societies today. They are:

*Keep moving.* The vast majority of food-related illnesses stem from a profound shift in human lifestyle, from constant, challenging physical and mental activity to sedentary life, punctuated only (if at all) by spurts of frantic exercise. Paradoxically, although a common belief is that our ancestors burned a lot more calories doing physical activity than people today, the evidence does not support this assumption. Moreover, although physical exercise has exploded in popularity in recent decades, obesity rates in North America also rose during that period. Critics of the link between physical exercise and obesity have pointed out that exercise just makes people hungrier, leading them to eat more, and the body compensates by lowering metabolism, obliterating any gains from working out.

As it turns out, the clincher may lie in patterns—rather than amount—of physical activity. (Decreased mental activity may also be a health risk, as discussed later.) Our male ancestors covered around nine miles per day on foot, while women walked around six miles each day. Walking decreases IGF-1 (insulin-like growth factor-1) hormone levels, which implies lower risk of hormone-related conditions like breast and prostate cancer and acne. Prolonged walking also reduces the risk of diabetes and is associated with weight loss. By comparison, the average American covers about two and a half miles on foot each day and spends the rest of the time sitting, driving, and watching nearly five hours of television. Watching TV is associated with an increased risk of obesity, type 2 diabetes, cardiovascular disease, and simply outright dying.

*Eat less meat and dairy when younger. Avoid sugar and deep-fried foods.* Meat was an important part of the human diet starting from at least two million years ago, but for most of the last twelve thousand years, almost all humans have had much less access to it. Cows and goats were milked by some populations beginning around nine thousand years ago, but for most of these populations, dairy was a much-appreciated *substitute* for meat and animal fat. Meat and milk are extremely nutritious, providing all the required amino acids, minerals, fatty acids, and vitamins to sustain human life, but the combination of these two nutritionally rich items in recent times was unprecedented in human history—and too much. Furthermore, as nations became richer, people were also able to purchase sugar and vegetable oils.

Hormonal activity, such as that of insulin and IGF-1, goes haywire when we consume a lot of animal protein and sugary and fried foods.[2] Eating more animal foods will probably make you taller, stronger, more fertile, and feel better, but these foods will also likely shorten your life span. Meat is most hazardous for the long-term health of younger people, because cancers and other chronic conditions generally take decades to develop and manifest. On the other hand, older people's health may benefit from the rich nutritional properties of animal foods. The irony is that we spoil our children by letting them eat liberally and then counsel ourselves to restraint when we are in our advanced years and suffering from chronic diseases. This is exactly the wrong strategy. If helping kids avoid chronic diseases is the goal, parents should limit their children's intake of animal foods; on the other hand, we can encourage elders to eat meat and be merry.

*Eat traditionally.* Rather than worry about what to eat and what to avoid, a good fallback is to eat traditionally. Traditional diets took centuries to develop and are based on how well combinations of food supported health and on how good ingredients tasted together. Faced with depleting meat stocks, our ancestors devised innumerable methods of cooking delicious and healthy meals that balanced nutrients. Furthermore, in societies where people lived on particular diets for hundreds or thousands of years, their bodies gradually became adapted to the diets, such as the case of lactose (milk sugar) digestion for people in northern Europe and India, and in East Africa. Eat traditionally and eat what *your* ancestors ate.

One problem with this strategy is that many people don't know what their ancestors ate. In *100 Million Years of Food*, we'll venture through the major stages of food history, from fruits, meat, starches, alcohol, and dairy to aquaculture and genetically modified crops, explaining why we ate various foods and the impact these foods have on our health. First, though, we'll step back a hundred million years in time and consider a food that some of our most distant ancestors evidently relished: insects and their creepy-crawly kin. Good enough for our ancestors—and good enough for us?

# THE IRONY OF INSECTS

The supreme irony is that all over the world monies worth billions of rupees are spent every year to save crops . . . by killing a food source (insects) that may contain up to 75% of high quality animal protein.

—M. Premalatha et al., *"Energy-Efficient Food Production to Reduce Global Warming and Ecodegradation: The Use of Edible Insects"*

*If you eat that ant, I'll never kiss you again.*

—*Ex-girlfriend during camping trip*

Few books advocating ancestral diets bother to mention bugs. This is rather strange, because bugs were once a major source of calories for human societies. It's a shame that so many people in industrialized societies today are repulsed by eating insects, because insects can be an excellent source of nutrition and eminently eco-friendly as well. On the other hand, insect cuisine has its drawbacks, which enthusiasts tend to gloss over. To understand what eating insects represents to our species, it's illuminating to step back in history, to a time when our ancestors licked their narrow muzzles in eager anticipation of a bug feast.

If you and I had been born 100 million years ago, we would have leapt from tree limb to tree limb in the depths of a humid tropical forest, scouring the leafy shadows for our favorite foods: skittery bugs that yielded a satisfying squirt of fat and peanutty protein when eaten.[1] There is some debate about whether our insectivore ancestors lived in Southeast Asia or Africa,[2] but during a recent visit to Vietnam, my inner early primate had me longing for grubs, so . . .

I post a notice on a social networking site for travelers: "Hello, I'd like to try eating insects at a restaurant in Saigon. Does anyone know of any places that serve insects or want to join me?" Within hours I have collected suggestions for restaurants that serve insects, and even a handful of intrepid dining companions. I scribble down contact information in my little brown notebook and head out the hotel door.

Phung Vong, a student with a willowy frame and gentle demeanor, is waiting on her motorbike, smiling at me. I sit up front and take the handlebars. Following Phung's directions, we zip along dark streets lined with massive fig trees, then over a long low bridge, a metaphorical crossing to another world. Almost immediately, we are lost. We drive back and forth along a broad avenue flanked by open-air beer and barbecue joints, searching for any signs of an insect eatery—but what would give it away? A peculiar aroma? Grasshopper legs dangling from patrons' mouths? We finally spot a large green and white picture of a jaunty cricket backlit by fluorescent white tubes: REC REC reads the sign, an allusion to the sound emitted by a cricket, in Vietnamese. By this point I am ravenous and ready to feast, even if it means tucking into a pile of jagged limbs, wispy antennae, and splayed wings.

Our fellow diners arrive on another motorbike; it turns out they were also lost. I'm curious to see what type of people gravitate to an open request for eating insects. We are joined by Nhat, a soft-spoken tourism company employee, and Andi, an Indonesian backpacker with flowing beetle-black locks. Nhat and I select live palm worms in fish sauce from the menu—seventy-five cents for each one.

Andi shakes his head vigorously when asked if he will eat insects. "Just chicken for me!" He met Nhat a few hours earlier through the Internet and has tagged along for the visual experience—he extracts a hefty DSLR camera from his bag—not the culinary experience. The two palm worms are brought in separate bowls, still alive, wriggling fiercely in a bath of turpentine-colored fish sauce with a few slivers of chili. The glossy brown heads of the grubs, the larvae of a weevil that infests palm trees, glisten like popcorn seeds; the wriggling abdomens have pale rubbery ridges. The owner of the restaurant, chubby and affable, comes out to instruct Nhat and me: we are to grasp the heads, pull off the fat white bodies with our teeth, and discard the heads, taking care that the larvae do not nip our tongues with their formidable pincers in the process. Biting down on squirming larvae seems barbaric, but my brain is starting to swim due to hunger, and the fish sauce is muskily aromatic. How bad could their fat glistening bodies taste? And am I not a direct descendant of insectivores, albeit roughly 100 million years removed?

I pick up my grub with chopsticks, examine the wriggling form at the end, then seize it carefully between my front teeth, taking care to keep the gnashing mandibles away from my tongue. A quick yank, and then my mouth tingles with a creamy infusion, pleasing in a mayonnaise-y way, especially in my famished state. I drop the decapitated head back into the bowl of fish sauce.

"Mmm! That was good!" I announce to the table.

Andi regards me with horror and newfound admiration.

Nhat's turn. Her hands wobble as she picks up the wriggling larva in her chopsticks. She puts it back down in the sauce, picks it up again. Breathes in deeply, out, in, closes her eyes, opens them again. The chopsticks quiver like guitar strings, but they travel upward, arduously, eventually delivering the squirming worm coated in rank-smelling fish sauce into her mouth, her teeth closing gingerly over the silk-white flesh just behind the bulbous head. Now it's my turn to admire Nhat.

The next dish is a monstrous fried centipede. It's all mine. No one else shows any appetite for it, not even brave Nhat. But why? If previous customers had fallen sick or died after eating it, the restaurant would have banished the centipede from the menu, right? I pick up the centipede by the tail. Although thoroughly fried, it sways back and forth in my hands like a wooden snake. If one squints one's eyes, the centipede could pass for a long, skinny tiger prawn. But my appetite and resolve are rapidly ebbing.

I bite into the centipede. The head is not bad, not much worse than eating shrimp skin. The middle section, however, is utterly vile and bitter. It is hard to resist spitting the whole thing out. No wonder these creatures slither about the forest floor with such impunity.

Insects were once a significant part of the diet in Southeast Asia, but under the influence of the French and Americans, they were increasingly omitted from Vietnamese cuisine. In Saigon, a megacity teeming with 7 million inhabitants, I can find only two restaurants whose menus feature insects and their ilk, such as scorpions and centipedes. What does this say about insect cuisine? Are insects such miserable fare that only poor people, dim-witted anthropologists, and daredevils eat them?

Thailand was never completely overrun by a European empire, in part because its rulers were able to exploit a rivalry between the English in Burma and the Malay Peninsula and the French in Indochina and thereby stave off invasion. As a result, the Thais were more successful at retaining their traditions, including insect cuisine. To see how insects are used in the context of a meal, rather than as novelty snack items at a beer house, I book a ticket to Thailand.

A few weeks after my Saigon insect excursion, I arrive late at night in Bangkok. I head by train to the Nasa Vegas Hotel, an imposing warehouselike structure on the outskirts of the city, conveniently located by the Metro line and priced within my modest budget. Since I have arrived on a Friday night, all the cheapest rooms are taken, but luckily I can afford to splurge for a single night in an Executive Deluxe room, exactly the same size as the budget rooms, but with swankier bedspreads and more solid doors. It's a good thing I don't believe in ghosts, because the Nasa Vegas is a ghost-friendly establishment: murky, tunnellike hallways, layers of dust, long sweeping staircases that few bother to ascend or descend except me. At four in the morning, I'm awakened by the sound of women laughing shrilly in the hallways.

The next evening, rain and wind lash the streets. I'm standing on the edge of a Bangkok boulevard on a Saturday night. I don't speak ten words of Thai, but I've always understood that if you keep your head up, smile, and look on the bright side, things are bound to turn out well. While tourists bumble about Bangkok in search of boobs, booze, and bargains, I slip into a convenience store. "Where can I find the bus to Khao San?"

The clerk and the people on the street offer conflicting directions—stand here, go there, look for this or that bus. A thin, pale, fashionable young woman notices my bewilderment. "Come with me!" she says, introducing herself as Milk. I obediently fall in behind her like a pup clinging to the heels of its mother. Milk says she was heading to Khao San anyway and offers to guide me to my destination. We board a bus, disembark at a mall, find one of Milk's friends in the crowd—another thin, sharp-featured girl with heavy makeup—then catch another bus to the tourist ghetto of Khao San. We thread through the drunken tourists and six-foot transgenders as a girl in a bar belts Adele's "Someone Like You" with enough angst to rip a heart to shreds. Across the passageway, adolescent girls twirl about on metal poles, their expressions distracted. Milk leads me through the throngs to a cart with a display of glistening fried insects. The seller is dark, plump, dressed in the plain manner of a country woman. Milk is proud of the bamboo-worm caterpillars. "I really like these! I used to eat these a lot when I was young," she exclaims.

The stringy white caterpillars could pass for chicken jerky in texture and taste. There is something incongruous about Milk, the paragon of hip Thai youth in her blue jeans, shimmery blouse, and brilliant makeup, tucking into a cup of fried caterpillars. Is this what the historians mean when they say that Thailand charted a course between the demands of the West and her own proud heritage?

I shell out some Thai baht for giant water bugs and black water beetles, each one about an inch long. Giant water bugs and water beetles were among my favorite pets during my childhood, when I collected them from puddles and ponds and raised them in bottles and aquariums. Eating and swallowing them, even after a good chew, is another matter. There are a lot of sharp edges to contend with. I feel like I'm snacking on disposable razor blades.

Milk takes me to meet two of her male friends, who order spicy green curry and fried noodles with pork and chicken from a stand. I watch the men hungrily shovel food down. Perhaps the insect carts in Khao San are a novelty after all, just like the palm worms and centipedes in Saigon. Milk and her friends look like they are on a happy double date. They say they will party and drink a lot that night. Feeling out of place and tired, I excuse myself from the group. Milk leads me to a taxi stand and tells the driver to take me to the Nasa Vegas. As the taxi pulls away, I wave, but she has already turned to rejoin her friends, the memory of the insect-eating tourist perhaps already fading from her thoughts.

The next day, I take the Metro, setting off around five in the afternoon. The girl next to me moves to another seat. I can't blame her. My shirt smells horrid because I tried to wash and dry my laundry in the room—the hotel laundry services were exorbitantly priced, with a load of laundry costing the same as a night's stay. Arriving at the vast Chatuchak Market, I wander along the stalls, the shops shuttering up in the late evening as rain sprinkles down. I find two little carts with some shriveled crickets and grasshoppers and not many customers. I eat a fried egg perched on a heap of noodles, sitting alone in the dark on a curb, feeling dispirited.

I read that insects are popular in northeastern Thailand, so a few days later I take a quick flight to the Thai–Laos border. My hotel room is airy, clean, cheap, quiet—a world away from the smog of Bangkok. Udon Thani is transected by narrow lanes traversed by trucks, cars, motorbikes, and elementary and college students. The local cuisine is sour, bitter, dazzling hot. By a stroke of good luck, while I'm struggling to communicate my inquiries about food in a market, a Thai American passerby rescues me from my linguistic abyss. Amy, who grew up in California, is in Udon Thani to start an educational consulting business. I tell her about my quest for insect foods. She offers to take me to Rajabhat University to find people who can assist me with finding local insect cuisine. With the help of Amy and staff and professors at the university, I end up with the cell phone number of a Vietnamese student who is doing his master's degree at the university and is fluent in Thai. He and I arrange to meet that evening in front of the university gates.

"Elder Brother Stephen!" Seven P.M. sharp, a young man pulls up at the university gates on a beat-up motorbike. He has a narrow face that is dashing in a daredevil way. A current of nervous energy seems to run from his toes to his fingertips. He hands me a helmet. "I borrowed the bike from a friend."

Hoang, from the hardscrabble north-central region of Vietnam, was sponsored by his seafood export company to study in Thailand. He plays tennis and teaches Vietnamese. When his face creases into a soft smile while taking a call from a Thai student, I guess that the local girls must flock around him.

Hoang ferries us to a gleaming downtown night bazaar. We come across stalls with lavish displays of glossy black water beetles, giant water bugs with wicked pincers, several species of grasshoppers and crickets, ants and pupae. The insects are fried in vegetable oil and seasoned with soy sauce. The grasshoppers have a solid crunch to them. The spurs on the legs take some getting used to. The mole crickets, with stubby arms, are lightly chewy with a tang. My favorite: ants, pleasantly soft with a bit of sour. You have to shovel down dozens to even make a dent in your appetite. When we've had our fill of soy-seasoned crispy limbs and papery chewy bodies—50 baht ($1.60) for a plate of ants, crickets, and grasshoppers—Hoang folds the remaining insects into a napkin, to take home to his friends.

During a morning stroll near the hotel, I chance upon bright red-orange papaya at a roadside cart run by a shy teenager. Her uncle, Mr. Amnat, comes out to chat with me and practice his English. He owns the laundry shop just behind the cart.

"I'm looking for a cricket farm. Do you know where I could find them?" I ask him. Seeing his confusion, I sketch a cricket in my note-book. Amazingly, Mr. Amnat says he knows just the place.

I return two days later to the papaya stall. Mr. Amnat drives up in a muscular black pickup truck. We surge past small markets and tight intersections. Mr. Amnat relates that he used to run a sugarcane farm, then began rearing pigs before settling down to the laundry business. I find it a little ironic that we are looking for eco-friendly insect cuisine in a gas-guzzling truck, but Mr. Amnat says his vehicle runs on 20 percent ethanol. We pop into a market where a woman sells two types of crick-ets, one light brown and crispy—preferred by Mr. Amnat—and the other a dark wine color and softer in texture. The cricket seller gives us directions to one of her suppliers a few miles out of town.

Out on the highway, rain begins to thrash the window. Mr. Amnat swings the truck into the driveway of a rural dwelling. Next door, some teens are partying with loud music, food, and drink. A bare-chested, potbellied, bespectacled man greets us. The cricket farmer, a retired municipal officer, invites us to the back of his house. Under a tin roof are fifteen or so concrete bins covered with blue mesh. Lively screeching fills the air. Peering through the mesh, I see thousands of plump brown-black crickets crawling over egg cartons and strands of vegetation. Pans of sand are placed inside for the crickets to lay their eggs. The farmer also shows us two pink-speckled lizards with huge eyes and suckered toes, geckos that he is raising in a dark wooden box for their meat.

Considering that crickets produce 50 percent less carbon dioxide than cattle per unit of weight gain and convert feed into food twice as efficiently as chickens, four times more efficiently than pigs, and twelve times more efficiently than cattle, insects deserve to be more popular on menus.[3] Since insects aren't warm-blooded, they don't need to con-sume as many calories as warm-blooded animals when putting on weight. Insects also use up less water than livestock per unit weight of flesh. A backyard operation like this farm, located in a warm climate (insects are small creatures and therefore are more vulnerable to cold than mammals) could contribute impressive quantities of protein for a surging, hungry population, yet the farm could still be readily managed by a retiree. It's hard to deny that edible insects could create a much smaller environmental footprint than equivalent-sized portions of meat, especially in densely populated countries that don't have space for rearing bigger livestock.


To help me make sense of insect cuisine, I return to Bangkok and catch a bus out to the suburbs, then wander across the carefully cropped grounds of Mahidol University, desolate on a Friday afternoon like many campuses around the world. A kindly, bespectacled woman is waiting for me in the Nutrition Institute. Professor Jintana Yhoung-Aree hands me a stack of documents on insect nutrition as she leads me out to her tidy compact car. It's a short drive to a busy restaurant that the professor says is famous for northeastern Thai cuisine. She orders fried frog legs with cabbage and bitter melon; fried chicken; bamboo shoots and baked spicy fish; coconut shakes with ice and sugar; and weaver ants and pupae in a fiery sour broth, flavored with lemongrass, mushrooms, garlic, and chili. Though my wimpy tongue can barely handle the ant soup, overall it's a brilliant, satisfying meal. Here, I finally taste insects in the setting of a varied cuisine, not merely deep-fried and served as street food, but as the accompaniment to a savory, balanced meal.

We are no longer pure insectivores, and our bodies have adapted accordingly. Chitin, the chief component of insect exoskeletons, is structurally similar to cellulose and may be a useful source of fiber, but primates that live largely on insects possess enzymes to digest chitin. Humans possess these chitin-breaking enzymes in our stomach juices to a limited degree, which means we are unable to extract some of the available calories from insect foods.[4] Eating insects has other potential drawbacks as well. Although insects are more dissimilar to us than mammals and birds and therefore may carry fewer of the deadly diseases that infected livestock can pass to us, raw insects can still transmit bacteria and parasites and thus need to be well cooked. Since insects often defend themselves by producing toxins with the help of plants they eat or may ingest pesticides or heavy metals spread by human industry, there's the possibility that insect toxins could seriously spoil a meal. Insect parts are also potential allergens, containing proteins in common with known allergenic animals like shrimp, lobster, and dust mites.[5]

Nonetheless, as side dishes, like the refreshing weaver-ant soup the professor treated me to, insects are perfectly admissible. After all, more than 1,600 different species are known to be eaten worldwide, which could not have come about if insects were harmful to health. The likely historical epicenter of insect eating was in the Americas, due to the fact that large herbivores were never domesticated in pre-Hispanic times and thus there was a shortage of protein.[6] Edible insects are practical in developing countries where meat is scarce or expensive, because insects provide essential amino acids, omega-3 and omega-6 fatty acids, B vitamins, beta-carotene, vitamin E, calcium, iron, and magnesium, sometimes in concentrations exceeding that of familiar meats like beef, pork, and chicken. The spurs on locust legs may snag briefly in the throat, and roasted silkworm larvae might not rival caviar anytime soon, but fried cicadas are astonishingly light and buttery, and queen termites are fabulous wrapped in egg. Moreover, having an interest in insect cuisine doesn't necessarily banish you to the realms of social doom—I have met many people who were willing to try eating insects during tropical adventures. It turns out that you can have your insects and eat them too.

Our earliest primate ancestors were happy to hunt down insects, which, to our diminutive forebears, were like triple burgers with all the fixings. However, the climate started to cool down, moisture levels increased, and the dominant tree types shifted dramatically. A new kind of tree offering a new kind of food emerged on the scene. Long before meat-eating became fashionable among primates, fruits arose as a tantalizing source of calories and nutrition, packing enough fuel to power the evolution of a new kind of primate, bigger and smarter than its insect-crunching forebears.

# THE GAMES FRUITS PLAY

Heroes on earth once lived, men good and great,
Acorns their food,—thus fed they flourished,
And equalled in their age the long lived oak.

—FREDERICK EDWARD HULME, *Bards and Blossoms;*
*or, The Poetry, History and Associations of Flowers*

*When durian falls, sarongs rise.*

—*Indonesian/Malay saying*

If the joys of eating are comparable to the joys of sex, as is sometimes claimed, then fruits—fun, attractive, breezy, and noncommittal—qualify as a summer fling. (By contrast, starches and vegetables are the in-laws, indispensable but tricky to deal with, sniping at us through a fog of flatulence and indigestion.) Who among us mortals would want to dispense with the teasing seductions of fruits?

However, the call of fruits is bundled in a sphinxlike contradiction. Across human history, fruits have generally served as accompaniments to meals, rather than the main course. Despite the seasonal availability and enticements of fruits, omnivorous animals such as bears and birds prefer to mix fruits with protein sources like insects and other prey. Bears and birds fed fruit-rich diets rapidly lose weight.[1] Humans also lose weight when their diets include large quantities of fruits.[2] Not a great way of losing weight, though: High concentrations of fructose, the predominant sugar in fruits, have been associated with excessive production of lipids, insulin resistance, pancreatic cancer, elevated uric acid levels, gout, cardiovascular diseases, and other metabolic disorders. Bloggers have speculated that Apple founder Steve Jobs's pancreatic cancer was related to his experimentation with extreme fruit regimens. Ashton Kutcher, an actor assigned to play Steve Jobs in a recent movie, was hospitalized for insulin and pancreatic issues after mimicking Jobs's fruitarian diet for a month to prepare for his movie role.[3]

In the early 1980s, a fifty-five-year-old farmer was admitted to a hospital in Toulouse, France, complaining of chest pains. The results of a preliminary examination were inconclusive, but an x-ray picked up a scattering of tiny nodules throughout the chest. The farmer suffered a fatal heart attack, and an autopsy was duly carried out. The examining physicians noticed a profusion of crystallized fatty acids in the victim's lungs. Analysis of the granules unveiled the presence of chemical compounds (hydrocarbons) commonly found in apple peels. When questioned, the farmer's family recounted how the farmer had eaten *a kilogram of apples every day for eighteen years*, amounting to perhaps five or six tons of apples over his lifetime. Though the investigating physicians believed that the heart attack occurred due to plaque accumulation in the arteries rather than apple consumption, in their report they remarked upon the striking manifestation of the lipid crystals throughout the victim's lungs.[4]

That an act as innocent as eating fruit could have egregious effects on the human body strikes the Westerner, raised on the virtues of "an apple a day keeps the doctor away," as bizarre. However, few people in traditional societies would think to stuff themselves on fruits. But why is this the case? Aren't fruits supposed to be the healthy food par excellence?

To tackle this paradox, let us first note that around 60 million years ago, our primate ancestors lost the ability to synthesize vitamin C.[5] Considering how extremely important vitamin C is to our bodies—it shields cells from oxidation, prevents scurvy, and provides critical amino acid and neurological (neurotransmitter) functions—ditching vitamin C is akin to a rock band firing the drummer. The show can go on, but why take such a drastic step?

There have been other cases of vanishing vitamin C synthesis. Ray-finned fish, a group that comprises 95 percent of the fishes living today, lost the ability to make vitamin C between 210 million and 200 million years ago, while close relatives like lampreys, sharks, rays, sturgeons, and lungfish retain the ability. Guinea pigs relinquished vitamin C production 14 million years ago. Bats ceded the ability to make vitamin C starting around 60 million years ago.[6] Many birds in the Passeriformes family, such as swallows and martins, have also lost the ability to produce vitamin C, while other birds in the same family, such as crows and mynahs, still have or regained that ability. Among our primate cousins, monkeys and apes cannot produce vitamin C, but more distant relations like lemurs and lorises still can.[7]

Remarkably, in all of these cases, only one gene was affected: the GLO (L-gulono-gamma-lactone oxidase) gene, which produces an enzyme involved in the last step of synthesizing vitamin C. When this gene is knocked out, *only* vitamin C production is halted. If other genes producing other enzymes affecting vitamin C had been affected, this would have had much broader harmful effects, and the organism would have been unable to survive or reproduce effectively. As it turned out, knocking out only vitamin C was possible during evolution because vitamin C could be obtained in the diet. Each of the vitamin C–deficient species listed above had a rich source of vitamin C available in what they ate from plant foods, insects, and so on. In these cases, making vitamin C was superfluous to survival, because there was already enough vitamin C to handle the animal's basic needs.

The loss of vitamin C synthesis around 60 million years ago implies that our ancestors must have had access to a lot of fruit or insects in rainforest environments. By 30 million years ago, our ancestors had evolved into frugivores, animals that subsist on fruits. Our molar teeth lost the sharp, narrow ridges that insectivores use to grind up chitin into edible bits and evolved instead the blunt molar teeth characteristic of committed frugivores. Though insects and leaves still needed to be eaten to supplement the protein shortfall of a fruit-only diet, the period from roughly 60 million to 30 million years ago marked the height of a heady romance between fruit-bearing trees and our lineage. Like flashbacks to a high school flame, the vestiges of this passion still smolder within us today.

Consider the case of durian, one of the world's most infamous fruits. Although renowned for its aphrodisiac qualities, it is also one of the most foul-smelling fruits on earth. Because of this, it is banned from public places and public transport in some Southeast Asian countries, even though it remains popular there. I remember vividly the time I lost my durian virginity—I was only twenty-five—under the tutelage of a barmaid who worked evening shifts at the Hard Rock Café in Saigon. Thin and restless, Tham had a habit of teasing me relentlessly. On her motorbike, we picked up a bulky, spiky fruit from a street-side vendor and conveyed the cargo to her friend's house. The women squatted around the durian on a tiled floor. A foot-long knife was procured. A breathless ambience seethed in the room. The women were excited—but at what? On the floor lolled a fruit reminiscent of an explosive naval mine, with a fearsome camouflage-green spiky rind. I sat next to the women, who ignored me, intent on this wretch of a tropical fruit. Tham's friend hefted the knife, thrust it into the durian, and carved out a hunk. Greasy golden fruit was extracted from the thick cavity, accompanied by a stench evocative of spoiled onions, rotten meat, and the menace of coal gas. The women dropped the pieces into their mouths, like hyenas feasting on antelope innards. I managed to eat about a fingernail's worth, then had to stop. It tasted as bad as it smelled, though the revolting creamy paste started to suffuse my body with tingly sensations. I felt like a frog being tapped with electrical wires. I wanted to spit out the durian, but that wouldn't have been polite.

In its natural habitat, cave fruit bats pollinate the durian flower in its lofty perch high above the ground. The bulkiness of the fruit means that small animals cannot be used to disperse the seeds. Therefore, the extraordinary smell is used to lure large mammals, who eat the fruit and distribute the seeds. The fruit also yields tryptophan, a precursor for the pleasure trigger serotonin, which helps to explain the fruit's reputation as an aphrodisiac.[8]

It is well known that fruits have struck a deal with animals: Fruits yield sugars, oils, or amino acids, and in exchange animals disperse the seeds far from the parent tree in their feces, giving the future seedlings a chance to conquer new territory. What is less well known is that many fruit plants may engage in a tricky love triangle.[9] The plants want their fruits to be consumed to transport their seeds, but bacteria, fungi, and insects also want to crash the party and feast on the succulent flesh. As a result, vulnerable fruits have to be protected with secondary compounds like phenols and tannins, which dampen the desirability of fruits to big eaters like ourselves by interfering with metabolism, digestion, and palatability but also restrain lecherous bacteria, fungi, insects, and other predators from spoiling the goods.[10]

One more consideration: Since plants want their seeds spread far away from the parent, plants may not be interested in fruit eaters who hang around the plant all day, monopolizing the fruits and discarding the seeds near the base of the plant. Like a mother who wants to wean her child or persuade her teenager to get a job and move out of the house, they must eventually practice some tough love. Plants can accomplish this by making their fruits enticing enough that eaters will gobble them down, but aversive enough that the eaters will eventually move on and leave the area. Thus the secondary compounds like phenols and tannins can be one more strategy for plants to ensure their progeny get broadcast far away from the parent. Or to put it another way, fruits were made to taste good, but not too good.

Humans take the eat-me-leave-me game yet one step further, by manipulating fruits so that their defensive compounds are neutralized. For example, olives are highly desired by animals for the high lipid content, but the fruits are also naturally bitter from their protective phenol compounds. People around the Mediterranean learned to tame these compounds through curing and fermentation. An extraordinarily hardy tree, olives were originally cultivated for oil that was valued as a means of lighting and as a skin lubricant, especially for ceremonial purposes (whence comes the term "Messiah," or the anointed one). Nowadays, the olive fruit is stripped of its bitter phenols and processed into various grades of edible oils with a high content of monounsaturated fatty acids, particularly oleic acid. (Phenols have excited interest for their potential role as antioxidants, but so far their health benefits remain unproven.[11]) The richness of olive oil complements foods like cereals, vegetables, fruits, and fish, by making dull but reputedly healthy fare more palatable. Hence the explosion in popularity of the Mediterranean diet: At last, Westerners can sit down to exquisite meals again without having to feel guilty and stressing out about calories and fat.[12] As with the phenols, there's little evidence that olive oil itself is a healthy food. What makes olive oil valuable is that it helps to bind together an entire regional cuisine, making it possible for people to subsist on fare that is relatively low in animal products like meat and dairy yet still feel reasonably satisfied, especially when fresh, high-quality olive oil is available and when people are too poor to buy meat.[13] As we'll discuss in a later chapter, humans are hardwired to crave meat because it increases our reproductive prospects; hence, when Greece began to increase her wealth after World War II, her citizens tended to give up the olive oil blessed by nutritionists in favor of the carnal pleasures of meat and animal fat.

Another major group of defensive plant compounds is tannins. Observant scientists have noticed that squirrels rotate their acorns so that the acorn "hats" point upward. Then the critters chomp through the caps and into the nuts. That's because tannin is concentrated in the bottom of the acorn, where it protects the seed embryo. (Irritants like urushiol and anacardic acid, related to poison ivy, similarly protect pistachio and cashew seeds, which explains why they have to be roasted before eating.) In high concentration, tannins render protein indigestible, inhibit a wide range of enzymes, sap energy, and stunt growth. Tannins are also sequestered in legumes, berries, and grapes, and they give red wine its characteristic dryness. (Incidentally, gray squirrels are better at digesting tannin than red squirrels and can therefore take advantage of oak tree forests. Red squirrels prefer hazelnuts, which contain less tannin than acorns, and will promptly kick the bucket if given nothing but acorns to munch on. The broader appetite of gray squirrels helps to explain why they are pushing out their red relations in England.) Birds like jays and grackles, as well as insects, similarly prefer to eat the top half of acorns. The rest of the acorn seems to have evolved to be a snack and lure for animals that do the necessary deed of widely dispersing the seed embryos.[14]

Oak trees can churn out roughly 500 to 1,000 pounds (225 to 450 kg) of acorns a year, albeit during a brief window of a few weeks. A Native American family living in California a few centuries ago, collecting over the span of two or three weeks, could set aside enough acorns to last two or three years. They could gather acorns from at least seven different species of oak trees, preferring oily acorns over sweet ones, and knew two methods to purge them of noxious tannins. The common technique was to de-hull the acorns, pound the acorn meat into mush and drop it into a pit, then douse the mush with water heated by hot stones until all the bitterness was leached. Alternatively, acorns could be buried in mud by streams or swamps for several months, after which they would become edible. To complement their protein-deficient acorn cuisine, Native Americans in California hunted salmon, deer, antelope, mountain sheep, and black bear and gathered earthworms, caterpillars (smoked and then boiled), grasshoppers (doused with salty water and roasted in earth pits), and bee and wasp larvae.[15]

The ancient Greeks made use of acorns; so did the Romans, as well as populations in medieval England, France, and Germany. Acorns were handy during the 1800s in Spain, Portugal, Arabia, Algeria, Italy, Greece, and Palestine and as of 1985 were still feeding people in South Korea, Morocco, and Iraq. Acorns are a cinch to gather, stay edible over extended storage, and provide a double whammy of fat and carbs, as well as vitamins A and C. These days, clever back-to-the-woods types stash their acorn meal in a cloth bag in the clean-water tank of a toilet, so that every flush expediently sucks away tannins. That marches the clever ones one step closer to ecological Valhalla, because not only do the oak trees lavish sustenance befitting a hero, but every call to nature effectively does double duty.[16]

So what's the catch? I had my first nibble of an acorn dish in South Korea, where acorn starch is boiled and set into a gelatinous block known as *dotorimuk*. From its lovely hues and quivering form, you guess that *dotorimuk* will taste like chocolate or almond Jell-O, but biting into an actual piece is like leaping into a kiddy pool at the height of a blazing summer: promising on the surface, with expectations soon dashed. *Dotorimuk*'s tannin aftertaste is reminiscent of boiled newsprint with a few peanuts tossed in, but Korean cooks mask this uninspiring flavor with a brilliant topping of scallions, garlic, red chili, sesame seeds, and soy sauce. The dish may have been useful during hard times in Korea, but given its unimpressive taste and the arduous process of boiling acorn paste through multiple changes of water, it is easy to see why this reliable commoner's staple has been displaced by more sophisticated fare. Professor Jared Diamond at the University of California, Los Angeles, has also pointed out that oak trees are hard to domesticate because they grow slowly, disperse seeds widely via squirrels, and have several genes controlling their bitterness.[17] Still, with all the talk today about the need for genetically modified (GM) foods and heavy fertilizer use to produce enough food for the world, it's sobering to think that our ancestors did rather well on ubiquitous fare like insects and acorns, which we literally tread on in our daily lives.

Acorns may be a poor person's panacea in the northern hemisphere, but what about the tropics? Is there some tree there that could also fulfill the needs of the hungry in an environmentally friendly manner? Perhaps breadfruit could vie for the title of tropical manna. I first sampled this remarkable fruit while traveling in Papua New Guinea, the region where it originated. A little smaller than the size of an American football, it is green and scaly to the touch outside, while the baked interior is golden and imbued with a nutty, starchy flavor that indeed recalls a freshly baked loaf. Breadfruit is typically roasted in an earthen oven with hot stones but can also be boiled and then used like potatoes. With a carbohydrate content comparable to cereal crops, breadfruit has the potential to be a super-productive tropical food machine. A single tree is capable of yielding up to seven hundred fruits in a single year, each fruit weighing two to eight pounds.

Though this glowing praise may make you eager to sink your teeth into a fruit that evokes the comforting flavors of baked bread or potatoes, cultivating breadfruit has its hurdles.[18] Breadfruit comes from a tall tree that can grow up to sixty-six feet in height, so much of the crop is damaged by falling and spoils within a few hours. (Breadfruit that is fermented in pits can still be eaten a year later, but decomposing breadfruit is an acquired taste.) The breadfruit tree does best in tropical regions with abundant rainfall and temperatures between 70°F and 90°F, so that rules out most Western countries, including the United States, where attempts to grow breadfruit in Florida have sputtered. The kind of breadfruit that is commonly eaten doesn't have seeds, so it must be propagated through root transplants, a substantial barrier to distribution. Early seafarers carefully nurtured their breadfruit cuttings on extraordinary voyages as they settled islands around the South Pacific.

Breadfruit trees take their sweet time in putting out fruit—around five years or so for the first crop—which may make them less profitable to grow commercially than, say, bananas, which bear their first bunches within one or two years.[19] Like a moody suburban teenager, breadfruit grows to lofty heights but takes a long time to mature, is picky about the weather and easily injured, and won't yield a penny on your investment for years to come. Fortunately, Hawaii residents Pierre Omidyar, founder of eBay, and his wife, Pam, have picked up the breadfruit craze and are bankrolling efforts to popularize the locally grown fruit in restaurants and with young people, which may help to cut Hawaii's extreme dependence on imported foods. The wood of breadfruit trees was once used to build houses and canoes (with breadfruit gum as caulking). The bark doubled as bedding and clothing, and the leaves served as trays for cooking and serving. The gum and bark were also used in traditional medicines, to relieve skin ailments, diarrhea, stomachaches, ear infections, and headaches.[20] The flower heads traditionally served as mosquito repellents when burned.[21] Thus, like acorns, breadfruit is a crop that has great nutritional, economic, and eco-friendly (as well as medicinal) potential, but it will require us to look both backward, to remember how people once used and ate this prodigious fruit, and forward, to devise technological workarounds to the problems of rearing, storing, and transporting breadfruit.

India is a paradise for lovers of fruits, flavors, and spice. I fly into the southwestern Indian state of Kerala, where my friend Bajish meets me at his parents' house on the outskirts of the city of Thrissur. He eyes me as I struggle to scoop up his mother's cooking with my inept right hand. Over the course of a week, she prepares breadfruit, chicken and fish curries, carrot and chickpea chilies, and tomato dahl, all steeped in rich coconut cream and tongue-flaying dosages of chili, accompanied by fried fish, eggplant, white rice, crumbly rolls of coconut and rice flour, crispy lentil flatbread, fragrant chapatti, creamy cassava, sour yogurt, and buttermilk. Bajish's expression is part amusement, part disgust. "You have to pick up the food like this," he says. To demonstrate, he puts his fingers together, like a spade, then raises them to his mouth with his elbow tilted outward. "Otherwise the food will fall down."

I try again, but somehow I just can't knit my fingers together and maneuver the grains of rice and bits of food into my mouth with Bajish's gracefulness. I have to keep my elbow lowered or the food spills, but then my fingers don't meet my mouth at the proper angle. It's a subtle art, but I have to master it, and soon, because the flavors of his mother's Keralan cooking have been astonishing.

Nostalgic Indian tunes blare from a little radio in his parents' bedroom. Bajish's mother swirls across the spotless white tiles, praying for her son's success. He's been looking for an academic job in oceanography for several months, ever since he graduated from studying in Japan, where I first met him. This next stage in Bajish's career is crucial, not only because it will provide him with status, income, and independence but because, according to his parents, it's high time for him, at the cusp of thirty, to get serious about marriage, and no good prospect could be contemplated without his having a decent career in place. So as my friend settles at the coffee table with his laptop, ready to send out more cover letters and résumés to universities around the world, his mother plies him with fried banana chips, papaya slices, and cups of ginger cardamom tea and wonders aloud about his marriage prospects. She is particularly intrigued by a girl in Thrissur who is educated and has a good office job.

Although Bajish is a gifted athlete and effortlessly acquired new strokes during swimming classes in Japan, he has been struggling to shed a few extra pounds over the years. At first glance, it seems nearly everyone around Kerala has a bit of a love handle or two around the waist. Nutritionists have blamed the high saturated-fat content of the traditional Keralan coconut-based diet as a factor in obesity and heart disease, but when Bajish and I chat with Dr. K. Jithendranath, an anesthesiologist with a strong interest in nutrition and traditional foods, he sputters at the attack upon coconut. "They said that coconut oil is bad for you, you get heart attacks. In that case all of my grandparents, everybody should have died! It's not the coconut oil, it's how they use the coconut oil. All that deep-fried stuff, browny-browny."

An analysis of the history of coconut diets and research on saturated fats supports Dr. Jithendranath's frustration. Coconuts are deemed objectionable by many nutritionists because of their high saturated-fat content. The unpopularity of saturated fats largely began after nutritionist Ancel Keys and his wife visited Naples in the 1950s. Dr. Keys's enthusiastic account of the Mediterranean diet enshrined it as the gold standard for the following generations. Keys and his followers argued that the diet of olive oil, whole-grain bread, fruits and vegetables, pasta, fish, legumes, nuts, and moderate portions of red wine, cheese, and meat led to lower rates of heart disease than the dreaded "Western diet." Subsequently, nutritionists trained in the Western sciences tended to laud the merits of the Mediterranean diet over other traditional and "fad" diets, such as the Atkins and Paleo diets, which are heavy in meat and saturated fats. However, critics noted that the statistical arguments of Keys's original papers were heavily flawed, relying on a cherry-picked group of seven countries to create an impression that lower consumption of saturated fat correlated with lower rates of cardiovascular disease. When the group of countries used to construct these arguments was broadened, the apparent correlation no longer existed. Since that time, the evidence linking saturated fats to heart disease has been tenuous and has not supported the stridency with which saturated fats have been vilified.[22] In fact, the particular kind of saturated fat found in coconuts, lauric acid, may actually increase "healthy cholesterol" (HDL) in the body.[23]

Moreover, in countries where coconuts were a staple in the traditional diet, chronic diseases only became prominent after Western foods and lifestyles were introduced (and coconuts phased out). For example, among the Tokelau Islanders in the South Pacific, the historical diet consisted primarily of coconuts, fish, and breadfruit. It was a high-fat diet: Over half of the calories came from fat, mainly saturated fat (roughly one-third of coconut flesh is saturated fat).[24] As the population of these atolls increased, the New Zealand government offered to resettle the Tokelau Islanders in New Zealand. About half of the Tokelau Islanders took up the offer and left for the mainland. Their new diet now included sugar, flour, bread, potatoes, meat, chicken, and dairy products. The result in the migrating population was an increase in obesity, type 2 diabetes, heart disease, gout, and osteoarthritis, even though fat intake actually *declined* after the move to New Zealand. On the other hand, daily sugar intake increased, along with carbohydrate and alcohol consumption. Among those who stayed on the Tokelau Islands, new European foodstuffs were also added to the diet, and rates of chronic diseases also increased, but not to the same degree as among the migrants.[25] Like olive oil for those living in the Mediterranean, coconuts make sense as part of a South Pacific or South Asian cuisine; the high fat content of coconuts complements lean fish and a largely vegetarian diet. Removing the anchoring effect of coconut invites dietary abuse in the form of novel fatty or oily substitutes such as fried foods, which are a known risk factor for diabetes and inflammatory disease. As will be discussed later, fried foods contain trans fats and AGEs (advanced glycation end products) and have a high glycemic index, which are novel and harmful characteristics in the human diet; coconuts contain saturated fat, a substance that our ancestors had moderate but steady exposure to over millions of years, mainly in the form of animal fats.

Bajish and I travel with a medical convoy into the hills around Kerala. The ethnic tribal people whom we chat with often use coconut in their diet, but obesity, heart disease, and type 2 diabetes are not medical issues among them. We also note their vigorous lifestyles, how they work the land with hoes and their hands and walk long distances to get around, in contrast to using motorcycles like the majority of Keralans. Kerala has among the best roads and highest income levels in India, but also the highest levels of type 2 diabetes. Petrol is heavily subsidized by the government, making it even easier to ride rather than walk. The risk of diabetes is strongly linked to a decrease in physical activity rather than to coconut in the diet.

Another key fruit of contemporary Keralan cuisine, chili, has also been viewed with suspicion by Western-trained nutritionists. The spiciness of chilies comes from the peppers' store of capsaicin, a chemical compound employed with excruciating effectiveness in pepper sprays (some spider venoms work through the same pain channel).[26] Chili plants seem admirably protected against predators, which might seem like a straightforward chapter out of plant evolution, but the saga of chili is wrapped in enigmas.[27] For starters, chili plants retain their pain-inducing capsaicin protection even after the fruits mature; most plants, by contrast, reduce toxins and make their fruits tasty at that stage, to invite animals to eat the fruits and spread the seeds widely. Also, not only have we humans come to enjoy chilies, many people seek out the wickedest varieties (*XXX!*, the hot sauce bottle labels trumpet, as if parading the temptations of adult entertainment). Why do humans enjoy the pain inflicted by chili-protecting capsaicins, the only mammals known to do so?

The most popular explanation of why we enjoy chilies is that their capsaicin compounds kill off fungal infection and other microbial invaders, and thus we come to enjoy chili dishes because we don't get sick from eating them. If this explanation is true, it would put chilies in the company of a long list of spices that humans use not only to perk up dull dishes but also to keep meats and sauces from spoiling (and people from throwing up and running to the toilet). When Paul Sherman, a biologist at Cornell University, and his then-student Jennifer Billing looked at spice usage from recipes around the world, they found that hotter countries used more spices. This makes sense, since increased temperature boosts bacterial growth and encourages food spoilage, thus making the need for spices more urgent. In particular, three powerful spices that inhibit many varieties of bacteria are more frequently called for in the dishes of warm regions. The knock-'em-dead spices? Most likely they are familiar to you and are tucked away in your kitchen cupboards right now: garlic, onion, and of course, our favorite sadomasochistic temptress, chili.[28]

However, there are some gaps in this explanation. The bacteria-busting hypothesis doesn't explain why chili seasoning is becoming popular in countries where food safety standards are high and food poisoning incidents are low, or why some countries that are geographically close to each other, such as Japan and Korea, have different levels of spiciness (Japanese food is considered relatively bland, whereas Koreans use chili in almost all of their dishes). If bacterial warfare were the only basis for eating spicy foods, then humans would get addicted to irradiated or canned foods, which seems not to be the case. The hypothesis also fails to explain why people steadily become addicted to eating chili, requiring ever greater amounts to feel satisfied. In fact, the more one looks at the behavioral pattern of chili consumption, the more it resembles thrill-seeking or recreational drug use.

Paul Rozin, a psychologist at the University of Pennsylvania, has suggested that humans are hardwired thrill-seekers, and we therefore enjoy blistering our tongues in the same way we (or some of us at least) savor a stomach-churning roller coaster ride and other forms of voluntary terror.[29] While equating roller coasters with chilies (and perhaps by extension garlic and onions and other spices) seems a little strange at first, back in the 1970s and 1980s, an American psychologist, Richard L. Solomon, pointed out that positive and negative emotions tend to come in pairs. When people are struck by lightning, survivors first experience terror, then elation. A similar thing happens to parachutists, who experience terror as they plunge through free fall; after landing, they warm up to a feeling of elation. People who take sauna baths go through an analogous sequence of discomfort followed by relief. The reverse is also true: When Solomon allowed babies to suck on a plastic nipple, they cried when the nipple was taken away. Solomon gave the article announcing his theory the clunky title of "The Opponent-Process Theory of Acquired Motivation," but fortunately he found a memorable subtitle: "The Costs of Pleasure and the Benefits of Pain"; that is, positive experiences are invariably followed by a drop in mood, while painful experiences are followed by relief.[30] Solomon argued that over time, the pain and the relief paired with it both diminish, so a person is compelled to repeat the experience with gathering intensity, resulting in addiction to mildly painful experiences.

Although psychologists today view the Solomon hypothesis as too simplistic to describe drug addiction behavior, it may help to explain the pleasure-pain paradox of spices. Ingesting chili is initially an aversive experience, but at small doses, the pain fades and a pleasurable state arises afterward. Other spices, perhaps many, have the characteristic of being initially distasteful but pleasing afterward. Not all aversive foods have these tendencies, though; for example, getting sick from food poisoning produces a prolonged period of nausea that no one wants to reexperience.[31]

We have addressed only one part of our original dilemma over spices. The second question remains: Why do humans alone come to enjoy mildly aversive experiences like chilies (and parachuting)? One possible answer is that humans are masters of gratification delay and brain rewiring. With practice, the discomfort of jumping out of an airplane, climbing onto a stage before crowds of thousands, or chomping on chilies gradually eases; however, so do the hits of pleasure, and thus the ever-increasing need for more punishment and more pain.

In other words, even though food that is spicy has antibacterial properties, we may eat these foods not to avoid getting sick, or even because they taste good at first, but primarily because they induce a paradoxical hit of pleasure after the displeasure; the benefits of pain,

Professor Solomon might have observed. One consequence of his theory is that it explains why tropical cuisines tend to be spicy: The lack of meat in them, especially fat, makes it necessary for cooks to drop in dollops of spices, to increase the feeling of pleasure that fat and meat would otherwise induce. When I lived in Korea, cooks who saw me about to ladle a spoonful of rice and vegetables without adding red chili paste cried out in horror, seized the nearest bottle of chili paste, and tried to squeeze it over my bowl, because they assumed that my meal would taste bland, but I had not been desensitized to chili by that point, so in my view, the pain did not merit the pleasure. Solomon's theory also helps explain why Japanese and Korean cuisines differ so much in their spiciness. As an isolated, fertile island surrounded by rich coastal waters, Japan historically had access to much higher levels of animal flesh, compared to peninsular Korea, and Japanese food therefore requires relatively small amounts of mustard (wasabi) compared to the full-force application of chili in Korean dishes. The same situation could apply to England, with its relatively spice-light and meat-heavy fare, and France, with its more flavorful but less meaty cuisine. The fact that spices inhibit bacteria would certainly have been helpful in promoting their adoption, but this may be an additional rather than the sole reason they're so widely used.

It seems logical that spicier, more flavorful food would make us fatter. However, chili may make people lose weight, by increasing metabolism, body temperature, and the burning of fat.[32] These weight-sloughing effects are modest unless chili is eaten in large doses, though, which limits its usefulness for populations unused to chilies, such as in the United States, Canada, and Europe. By contrast, in one Mexican study, the average person ate the capsaicin equivalent of seventeen jalapeño peppers *a day*. Unfortunately, there is some evidence that eating copious quantities of chili could increase the risk of stomach, liver, bladder, and pancreatic cancer. Scientists at Kyoto University have developed a new variety of chili, CH-19 Sweet, that could offer the health benefits of capsaicin without the pain.[33]

———

Between 40 million and 16 million years ago, something curious happened to our ancestors: Our uric acid levels started to rise because our ancestors progressively lost the genes for manufacturing uricase, the enzyme that helps dispose of it. Uric acid, a by-product of a diet rich in purines (organic compounds found in seafood and beer) and fructose (the sugar in fruit), can be a very inconvenient, nasty substance. It's responsible for causing gout, a debilitating condition in which crystals build up in a sufferer's joints. As a result of losing the ability to manufacture uricase, humans have uric acid levels three to ten times higher than other mammals and unfortunately a greater predisposition to gout and possibly hypertension. The loss of uricase over millions of years of evolution is one of the greatest unsolved mysteries in the evolution of the human diet. Because high uric acid levels are dangerous to health, it's extremely puzzling that our ancestors progressively shed the ability to deal with uric acid. Like losing a kidney or lung, it may not be fatal, but it's a considerable inconvenience. Why did our evolution take us down such a hazardous path? Around 70 percent of our uric acid is resorbed by our kidneys, not excreted, evidence that uric acid must have some positive role in the human body, rather than simply being a nuisance by-product of purine as scientists had formerly believed.

Many hypotheses regarding the function of uric acid have been proposed. One suggestion is that uric acid helped our primate ancestors store fat, particularly after eating fruit. It's true that consumption of fructose induces production of uric acid, and uric acid accentuates the fat-accumulating effects of fructose. Our ancestors, when they stumbled on fruiting trees, could gorge until their fat stores were pleasantly plump and then survive for a few weeks until the next bounty of fruit was available. The problem with this theory is that it does not explain why only primates have this peculiar trait of triggering fat storage via uric acid. After all, bears, squirrels, and other mammals store fat without using uric acid as a trigger.

Some researchers argue that the elevated levels of uric acid that accompany gout could have been a survival advantage in ancestral environments that were arid and where food was scarce, because high uric acid levels are associated with increased blood pressure (which is dangerously lowered when salt is scarce) and a greater tendency to deposit fat. Uric acid could have helped to maintain adequate blood pressure in a low-sodium fruit diet and during an interval when Earth's climate was drying out and hence salt loss through sweat could have been a problem.[34] However, mammals that thrive in arid environments, like camels and desert mice, seem to do fine without elevated uric acid levels.[35] Other mammals also subsist on fruits, but primates are the only animals known to have lost uricase. According to yet another hypothesis, primates are pretty smart creatures, and most of them lack uricase, so therefore uric acid must be responsible for their increased intelligence. While it's true that higher levels of uric acid have been found to protect against brain damage from Alzheimer's, Parkinson's, and multiple sclerosis, high uric acid unfortunately increases the risk of brain stroke and poor brain function.

Those trying to solve the mystery of this trait in human history try hard to recast symptoms of high uric acid as being beneficial in our past. This is a common tendency in evolutionary theorizing; people try to find an evolutionary reason in facts that may actually be by-products of evolution. The cognitive scientist Gary Marcus labeled such evolutionary by-products as "kluges"; some aspects of our bodies, like bad backs, arose because something else had evolved—walking upright, in the case of bad backs—and we humans got stuck with the accidents of history.[36]

A more realistic proposal for the evolution of uric acid has this character of kluginess. After several million years of not producing vitamin C in fruit-rich rainforest environments, our primate ancestors had no way of reevolving this ability because too many mutations had accumulated in the original vitamin C–synthesizing genes over the long period of disuse; like a car engine too long unused, vitamin C synthesis could no longer fire up. As it happens, uric acid has chemical properties that permit it to function as an antioxidant.[37] Therefore, the adoption of uric acid, a by-product of eating fruit and insects, was a possible second-best defense against oxidants. Indeed, higher levels of vitamin C result in lower levels of uric acid and diminished gout, possible evidence that vitamin C and uric acid are partial substitutes for each other.[38]

Like any evolutionary adaptation, there were drawbacks to uric acid's newfound role as an antioxidant. Exposure to high uric acid levels from overabundant fructose and purine consumption over several years results in insulin resistance, hypertension, and obesity-related disorders. In the ancestral environment, encountering significant quantities of either fructose or purine would have been rare. Today, fructose is plentiful, in the form of soft drinks and sweet, overdomesticated fruits like apples and oranges; purines are also common, found in seafood, meat, lentils, and other foods. A recent study also observed that high uric acid levels are associated with greater excitement-seeking and impulsivity, which the researchers noted may be linked to attention deficit hyperactivity disorder (ADHD).[39]

Blocking the production of uric acid through drugs like allopurinol alleviates hypertension, at least with adolescents who are not too far down the path of uric acid–mediated damage. However, drugs that reduce uric acid may cause serious side effects, such as immune system reactions resulting in fever, rash, impaired kidney functioning, liver damage, and elevated white blood cell counts.[40]

At this point, if you were to hand the script over to an imaginative sci-fi writer, he or she might suggest injecting people who suffer from high uric acid levels with uricases from nonprimate animals, or re-creating our ancestral uricase on a computer, synthesizing it in a lab, and injecting it into patients.

Truth is stranger than fiction: Researchers recently combined pig uricase, which is highly effective in breaking down uric acid, with baboon uricase, to lower the risk of immune rejection from human recipients. Although this pig-baboon chimera uricase was effective in reducing uric acid levels, it broke down very quickly in animal tissues and required chemical modification to become stable. Unfortunately, this modification also made the chimera uricase more likely to be rejected by human immune systems. Researchers then used computer programs to reconstruct a uricase that we last possessed *92 million years* ago. The ancient uricase was synthesized in a laboratory using handy

*E. coli* bacteria as surrogate mothers for the synthetic enzymes. When injected into healthy rats, the ancestral uricase was found to be a hundred times more stable than the chimera uricase, making it a promising candidate for drug development.[41]

To put everything into perspective, fruits, like insects, were once an integral part of our evolutionary history and remained a valuable part of traditional diets. Even though meat provides virtually all of the nutrition necessary for survival, at certain times fruit could be crucial to human health, especially when fresh meat and its accompanying vitamin C were unavailable. For example, the Inuit living in Alaska, northern Canada, and Greenland made use of a broad variety of animal foods—seal, whale, walrus, caribou, polar bear, fox, wolf, Arctic hare, waterfowl, fish, mussel, sea urchin, and so on—but the Inuit also harvested a staggering variety of berries. These berries were critical; Inuit who lacked fresh seal meat could develop pustules when the berry crops ominously failed, as occurred in 1904–5 among the Greenland Inuit.[42]

From being a seasonal snack in traditional settings, fruits in industrialized countries have become sweet, cheap, and holy: Fruits offer urbanites, weary of the associations of meat with disease and cruelty, the opportunity to detox with spiritually unblemished food, a karmic train that inches forward with every four or five bucks forked out for a mega-sized fruit smoothie. Sadly, our ancestors jumped off the tracks leading to Fruit Heaven 16 million years ago, rendering our genes and livers unsuitable for daily jug-loads of fructose. Such a dilemma, however, only arises when we lose sight of cuisine and obsess instead over nutrition. Unlike the here-today-gone-tomorrow wonders of scientific-pop nutrition, traditional cuisines are products of exquisite culture, symphonies of flavors and complementary foods that arose from the mistakes and insights of generations of eaters. People in traditional societies ate fruits in moderate quantities that their bodies could absorb.

Traditional cuisine, in turn, is intimately tied to ecology, the plants and animals that are naturally suited to a given place. Plants and animals can be grown and raised on industrial-sized operations that require a blizzard of chemicals and automation, but people around the world are experimenting with permaculture, the notion of living in an ecologically sustainable manner. Fruits and nuts are important in this movement because they could provide more food than meat could refurbish for a given plot of land. Soon after visiting my friend Bajish in Kerala, I discovered by chance that one of the most visionary and brave pioneers of permaculture lives in the Indian state of Goa, on the western coast of the country, not far from Kerala.

I lean into the turns around Chorao, an island tucked into the hip of the slow-flowing Mandovi River, and sweep past dense fronds on either side of the narrow road, the occasional motorbike or small truck over-taking me. I'm in no rush.

My new friend Hyacinth wrote out the directions last night at her dinner table in her architect's clear script, amid a medley of banana liquor, stir-fried squash, and bitter melon. A crazy kitten kept jumping on the table and kitchen counter, trying to pilfer food. The directions were to a farm owned by a young woman who was a client-turned-friend of Hyacinth. The next morning, after a breakfast of fragrant, warm homemade naan bread, Hyacinth and her friend Jean chaperoned me to a motorcycle rental shop on an oil-blackened side street. The two ladies haggled with the slicked-hair dealer over the rental price. Jean used her status as a high-ranking bureaucrat to vouch for my name, in lieu of leaving my passport. She whispered at me, in a fierce tone, "You can never trust these people, Stephen. Always keep your passport with you." I was always happy to know someone local; it was better than having insurance. I bade the ladies farewell, fired up the motor, and puttered out to the little highway running along the coastline. On my way out of the city, I stopped at a liquor store to buy a bottle of Indian rosé wine of uncertain quality (trusting the advice of the proprietress), then buzzed across the two-mile cause-way, just making it onto a ferry before it set out across the sluggish brown river like a hippopotamus.

I ride across Chorao Island, appreciating the scenes of dense foliage, quiet streets, scattered houses. Soon I regain the mainland.

Hyacinth's map is a montage of strong arrows, confident circles, bold lettering. The route is roundabout in comparison to the optimal line that Google Maps conjured, but now I appreciate that Hyacinth wanted me to savor the forests of Chorao Island, a welcome respite from the dusty wide roads and manic flow of trucks and motorbikes on the mainland. Two hours later, I arrive in a small village. My cell phone has no signal, so I ask to borrow a phone from the man behind the counter at a grocery store. I call the number that Hyacinth gave me, trying to reach the owner of the permaculture farm. No answer. I call Hyacinth, who is busy conducting exams for architecture students, and she tries the same number and also gets no answer.

I sit down for a lunch at a restaurant, politely turning down the offer of cutlery, because I'm starting to get the hang of scooping rice and curry with my fingers after patient coaching from Bajish. The curry is blistering hot, but after a week of digging into Indian cuisine, I've lost my extreme sensitivity to chili heat. The owner of the restaurant tries calling the number. No answer. The brother of the owner of the restaurant tries calling as well, but no answer. I call Hyacinth, and she, still in the middle of conducting exams, tries again, with no luck. The brother of the owner of the restaurant takes me to his roadside shop, which sells packaged snacks and drinks. He offers me a finger-length sweet banana. His wife, standing behind the counter of their shop, calls the number. No answer.

Hyacinth calls back on the restaurant owner's phone to tell me the name of the farm, which she learned from a colleague. The restaurant owner gets on my rented motorbike, and I hop on behind. After a few minutes, we take a side road that undulates through thin forests and scraggly fields, reaching a hand-painted sign that announces Foyt's Farm. We ride along a bumpy lane, fumble with a rickety gate, then come to a low red-tiled bungalow. I go around to the back, where a woman and man are sitting on benches and engaged in rapt discussion. Seeing me arrive, the man departs and the woman rises to greet me.

"Oh, you finally got here. I was wondering when you would get here," she says.

"I'm Stephen. I tried calling your phone . . ."

"Ah. I must have turned it off."

Her eyes sparkle and her chin is held high. However, she also carries an embattled air. Clea is the owner of Foyt's Farm, a twelve-acre working farm and learning center, and she never seems to stop moving. On the day that I arrive, she has been busy trying to preserve an insect-infested tree by coating its base with a natural pesticide, finding out what happened to all her chicks ("Perhaps a hawk or mongoose got at them"), and directing the workmen who are installing a sink ("It's crooked!") in her guesthouse bathroom. Clea has been offering instruction in permaculture and plans to take in more students, but for that to happen she needs to upgrade the bathroom facilities. While my host whirls about her tasks, I take a nap on an inviting outdoor bed, next to a tethered calf that was separated from its mother for medical treatment. Once in a while the calf belts out a loud *mooo* and is answered by its mother on the other side of the house.

The workers finish at five in the afternoon, and Clea comes over to chat with me, looking much more relaxed. Trained at Cambridge as a plant physiologist, she put her Ph.D. on hold to start a permaculture farm in her native country (the weather and food in England didn't suit her). With her father's help, she eventually bought an abandoned farm that she had fallen in love with, a dozen acres deep in the backwoods of Goa. Her self-stated mission is "I want to revolutionize the way that Indians conduct agriculture."

She and I set off on a tour of the farm. First it's the chicken coop, where twelve hens (sadly without their chicks) are kept at night to protect them from panthers and other nocturnal predators. Clea doesn't raise the chickens for their meat; their job is to pick off termites.

"All the wood structures are safe. Ask all the architects in Goa, they say you can't use wood: 'All the termites!' We have termite mounds around the farm, but no termites around the house. I think it's because of my chickens."

The chickens also poop in the soil and aerate the earth through their digging and scratching. Once in a while, Clea harvests the hens' eggs. We approach the cow stall where Clea keeps a few cows and their calves. Like the chickens, these cows have it pretty good; they will be neither milked nor butchered.

"I just want them for their piss and shit," she says, in her characteristic blunt manner.

Cattle urine and manure make excellent fertilizer. In true permaculture spirit, she also captures the waste from her own toilet to grow her vegetables. She discovered through experimentation a method of treating urine and feces from a regular flush toilet; the toilet output could be used to grow vegetables, with no adverse health effects noted.

We descend into the fruit terrace, which resembles a modern-day Garden of Eden: trees bearing juicy sweet lime-green star fruit, crunchy rose-pink heart-shaped love apples, and a profusion of cashew apples, odd yellow-red triangular fruit with hooked appendages harboring the cashew seed. While I'm stuffing my face on star fruit and love apples, we reach an old concrete dam and the waterfall that runs beneath it. Clea is trying to design her guesthouse bathroom so that slabs of cool smooth gray stone in the shower re-create the ambience of bathing in the waterfall. We ascend a short ridge, just as the last of the light fades from the warm Goan night sky. Clea points out a distant phalanx of trees demarcating the valley. In the middle of her property stands an old crooked tree with menacing spindly limbs raised against the sky, bristling like a gargoyle. "The Devil's tree," she says, laughingly.

According to the previous farmer and local folklore, the property was cursed by this tree and crop yields were consequently low. The farm was abandoned, and Clea was able to buy the property, one step closer to realizing her dream. Clea notes that the true value of her long scientific education was that it equipped her with a skeptical mind and a love of experimentation. The farmer's principal error, to her way of thinking, was that he used crude rice monoculture practices, stripping the land of its natural cover and exposing the nutrients to leaching and erosion. It was this practice, rather than the Devil's tree, that doomed his crops. Clea takes me to a patch of earth on the farm that is caked, dry, and barren, as if it were the boot-print of a giant. Clea says that when she bought the property, all the earth looked like this.

Over a span of four years, with the help of her students and hired work-ers, Clea has slowly brought the land back to life by analyzing the soil chemistry and water runoff patterns, using plants with matted roots and weeds to hold the soil, and employing a variety of other ingenious techniques. Foyt's Farm is an example of a self-sustaining ecosystem, whose general strategies can be used in other places to return to a more locally efficient method of farming.

In the remaining eerie dusk light, we gather fallen cashew fruit in our hands, the smell of the soft, broken fruit lustful and cloying. Clea planted trees on her property that give off strong perfumes, to con-fuse insect pests as they try to home in on the scent of their preferred host plants. This lessens the need for chemical pesticides and provides a natural fragrance.

The strenuous beating of drums can be heard in the distance.

"It's India. There're always festivals going on," she tells me.

I follow the thin swath lit by Clea's torch as we head back to her dwelling. Clea harvests arugula and lettuce from the greenhouse (rat-like bandicoots have been breaking in and wreaking havoc) next to her open-air kitchen. She pulls out a package of homegrown cashews from a freezer and pulverizes them into a cream with a countertop blender. One of her workers fetches a strand of black and green pep-per from a vine growing around a tree near the house. A squirt of homegrown lemon into the cashew cream, a sprinkle of fresh pepper-corns. We tuck into the meal. The heat of the peppercorns and aru-gula and sour lemon sets off the cashew cream perfectly. I ask for another helping. And another.

"Have some more dressing," Clea says. "Every day I eat this and every day I get excited!"

"You're living in paradise," I respond, overwhelmed by the feast.

"I think so, but no one else does. They think I'm crazy."

We happily eat, slurp, and grunt. Clea sighs, indicating her ample waist. "I eat far too much, though, way more than I should. I have a lot of healthy meals, but when the bananas are ripe, I eat six or seven a day. I eat cashews, all fattening!" Clea notes my empty bowl. "Shall I get you another star fruit? This is a meal!"

It's remarkable to think of what Clea has achieved on the strength of her determination, ingenuity, and passion for sustainable farming. By comparison, Clea's sister is the model of Indian respectability: educated at Berkeley in economics, then Harvard Business School, and now employed at one of the biggest management consultancies in the world. Clea's sister is married, has two kids and a big house in Belgium, and employs a maid. Clea expresses not the slightest iota of envy for her sister's life. Clea recalls that she never played with dolls; she once asked her parents for an elephant and horses but was given instead a dog, ducks, rabbits, tortoises—anything her parents thought she could handle.

I bring out the rosé wine. It is terrible—never trust these liquor store owners—but as we sit there in the utter darkness, Clea becomes expansive. Her parents wanted her to get married, but Clea's relationship is on the rocks.

"Who wants to live with a crazy woman in the jungle?" she says.

We listen for the sounds of a panther that roams the perimeters of the farm, stalking the chickens. Clea seems unperturbed. "You can sense its presence, the way a twig snaps . . . not like a wild boar blundering in the bush."

It's unnerving to think that a large predator is roaming beyond the circle of feeble light emitted from the house; I think I might have to pee soon. A tiger had been spotted before in a nearby village, and a twelve-foot cobra was caught on the property once. Clea says this is when she is happiest, alone at night, sharing the jungle with wild animals, soothed by the entrancing perfume emanating from her trees, living on a farm that had been abandoned but has now begun to heal, yielding the sweetest, most satisfying fruits that a human could possibly ask for.

# THE TEMPTATION OF MEAT

I shot a large roan antelope which was divided among twenty-two adults and forty-seven children in a community where there had not been much meat available recently. . . . [An] old lady cried light-heartedly, hitting her stomach, "I have been turned into a young girl, my heart is so light."

—Audrey Isabel Richards, *Land, Labour and Diet in Northern Rhodesia: An Economic Study of the Bemba Tribe*

Fruits have their devotees; in modern times, some people have subsisted largely or even entirely on fruit. However, in the long history of food, our affection for fruit pales in comparison to our worship of meat. It's true that some people abstain from meat, many of them repulsed by meat's taste and associations, but instances like this represent triumphs of mind over body. Human babies are not so complicated. A few days ago I watched my nine-month-old nephew in action during lunch. Arrayed on his plastic tray were diced bits of carrots, green beans, and stewed pork. His pudgy little hand reached out and vigorously swept aside the green beans, seizing instead the morsels of pork, every last one. In the years to come, he might ponder his relationship with meat and spurn it for a spell, as many of my friends and family have done, along with myself; for now, his genes have programmed him to target meat over veggies. In the turbulent religion of food, infants dwell peacefully, having no inkling of sin. The puzzle over meat goes beyond morality, though; even from a biological or archaeological viewpoint, many researchers grapple with why and how humans developed a taste for meat, the hallmark of the Paleo or caveman diet.

Dominic looks me up and down. A former rugby player twice my girth, he could quash my toothpick frame at a moment's notice. An overhead fan spins sluggishly, not enough to churn and dissipate the thick heat coiling around us and the other waiting passengers. "You're a student?" he asks. "You can visit my village, if you wish. I live in the rainforest. It won't be what you're used to." He steps out into the sun to buy me a Coke from a vending machine, sealing our agreement in a stream of ice-cold fizzy brown soda.

I'm in Papua New Guinea to scout a future site for my dissertation research. Before writing their dissertations, anthropology graduate students are required to undertake fieldwork. Papua New Guinea has long been a magnet for anthropologists, because the region's steep mountain ranges and thick jungle terrain led to the evolution of a profusion of cultures and more than eight hundred distinct languages.[1] I selected the province of West New Britain for my fieldwork site because from satellite and atlas maps, only a single spindly road could be seen worming into the interior, which meant that I would have a better chance of observing more intact cultural traditions. However, the lack of infrastructure in West New Britain also meant that it was impossible for me to arrange contacts in advance. To my great fortune, after I caught a lift to an airport on the northern side of West New Britain in the back of a pickup, my driver happened to see his friend Dominic in the waiting room and introduced us, giving me a precious contact in the interior.

An hour or so later, our bush plane shudders over a forested mountain range. The pilot boomerangs over a cove and drops us swiftly onto a grass landing strip. Dominic leads me through the dusty lanes of the town of Kandrian to stock up on tinned mackerel, rice, batteries, kerosene, and tarp. We lug the boxes to an outboard boat. When enough passengers are gathered, the craft skims out of the cove like a flying fish. We endure numbing sea spray, but after nightfall the waters are lit by glimmering bioluminescent plankton. Eventually we spot flickering torches and huts on stilts. Dominic's name is taken up by a chorus piping in the night. Villagers wade out to the boat and hoist our belongings to shore.

At daybreak, the outboard boat is maneuvered into the maw of a mangrove channel. We pole through shallows, then hike two hours along a slippery trail to the hedges of bougainvillea bordering Dominic's village. Dominic introduces me to his nephews Aloish and Frank, two easygoing New Guineans in their early twenties. Dominic quickly organizes a work party to build a hut for me next to his house. The walls consist of three panels of aluminum siding, with another piece as the roof. Branches are lashed together to make my bed, and two other beds are fashioned for Aloish and Frank. I hang my mosquito net over the bed. Aloish and Frank build up a fire in the middle of the hut. I have a thin blanket to keep me warm and a mosquito net to keep off the bugs, unlike the two young men shivering in the cold.

The Gimi people subsist on cassava, yams, grated coconut, and greens, with occasional windfalls of pork. One evening, my provisions are exhausted and only crumbly yams are doled out. Aloish and Frank request a few bills of Papua New Guinean kina from me to buy flashlight batteries. In the morning, I lift my mosquito net to find a tin pot hanging by the foot of my makeshift bed: a boiled fruit bat, eyelids wrinkled in resignation, rests in a creamy coconut broth, with some strands of bitter greens tossed in. Primal craving drives me to scarf down the bat, grayish oily skin and all, sparing only the bones and brains.

Cooks around the world, past and present, have recognized that great food depends on the glories of fatty meat or some other kind of fat. The Pacific Coast Native Canadians prepared prodigious quantities of oil from salmon and oolichan (candlefish) for their feasts. Coconut and sesame oils impart to South Asian and Korean dishes their comforting tastes. Lard was used throughout much of Eurasia to transform the meat-poor meals of peasants into proper fare, while whale blubber, beaver tail fat, sheep fat, kangaroo fat, whole milk, and olive oil were celebrated elsewhere. The desire to consume fat can humble a grown person. In Melbourne, Australia, a lady on a bus told me that her portly grandmother made stealthy forays into a bucket of forbidden but coveted lard, forging a pact with the grandchildren to keep her indulgence from her daughter's knowledge.

Yet there is nothing intrinsically nutritious about meat, fat, and oil; after all, gorillas, giraffes, and elephants are whopping big beasts that thrive on vegetarian diets. Some of the longest-lived peoples in the world achieved excellent health while consuming sweet potatoes, wheat, corn, or rice and very little meat.[2] Why are we so crazy about meat, fat, and oil? And is eating such fare good for our health?

To answer this question, we need to look at our family tree. As the Earth's temperature cooled down over the past 50 million years and rainforests became scarcer, our ancestors developed different niches. Our orangutan and gibbon ape relations spend their lives in the rain-forest canopies of Southeast Asia searching primarily for fruit. Another of our ape relations, *Paranthropus,* lived in Africa 3 million years ago and chomped on tough plant foods, aided by massive jaws and molar teeth that were perfect for grinding. *Paranthropus* skulls look similar to gorilla skulls. If we had evolved from *Paranthropus,* today we would all be content to munch on leaves, grasses, seeds, and roots, and the grounds of a North American college would be a smorgasbord for gorilla-head humans: lawns, leaves, flowers, acorns as a fallback snack. But students, no matter how skimpy their budgets, do not browse on the shrubs outside the dormitory, because humans did not descend from Paranthropine herbivores. Despite the apparent abundance of menu options, seasonal variation in the quality or availability of plants forces herbivores into long marches or airborne flights for lusher pastures.[3] The natural range of *Paranthropus* was likely restricted, as is the range of modern-day gorillas. The geographical options of an omnivore are considerably broader.

Our closest ancestors have a marked fondness for fresh flesh.

Bonobo apes supplement their fruit diet by snatching unwary small mammals like antelope, while chimpanzees use their intelligence to form hunting packs and trap and ambush their colobus monkey prey.[4] Though not as nimble as bonobos and chimps, orangutans and gorillas also take advantage of opportunities to hunt.[5] Our ancestors seem to have increasingly enjoyed the taste of meat. Scientists have unearthed cut-marks on bones and tools from 2.6 million years ago.[6] Genetic analysis of tapeworms, picked up by eating infected meat, suggests that humans, tapeworms, and meat eating went together as far back as 1.71 million to 780,000 years ago.[7]

But how exactly did our *Homo* ancestors obtain their meat, and how much of it were they able to get? This topic has been at the center of perhaps the most vociferous debate regarding the history of humankind, because the answer reveals whether you think the "natural" human behavioral pattern is hunter or scavenger. One can see immediately how this research has ramifications for today's quarrels about the merits of meat-heavy diets versus vegetarian diets.

The original hypothesis was that our early ancestors were first and foremost hunters. However, African predators typically rely on stalking and powerful bursts of speed, often hunting at night; by contrast, *Homo* was a daytime creature and could not sprint as rapidly as a lion or cheetah, nor did it have the claws, powerful jaws, or fearsome teeth of carnivores. How could any ape using sticks and stones hope to scratch, much less kill, a springbok antelope that could race down a savannah at 62 mph and catapult twelve feet high on a whim?

A popular counterproposal in past decades was that our ancestors filched bones and braincases of leftover game from under the sleepy gazes of sated lions and hyenas. The transformation of Man the Macho Hunter into Man the Bone-Heap Diver was also more politically correct. Since our ancestors were believed to be bargain-bin hunters rather than cold-blooded executioners, modern humans could be imagined as natural pacifists by extension. As an added bonus, the scavenger scenario was less sexist, because rather than being the family provider, Dad supplemented the main meal, Mom's plant-based stew, with marrow, brain, and scraps of meat. The scavenger proposal, alas, may not have enough meat to it. How was early *Homo* able to fuel a dramatic doubling in brain size in a million years on greens and tubers flavored with marrow and brain bouillon? Moreover, no mammal today lives entirely or even mainly from scavenging carcasses, because such opportunities are inherently unpredictable, meager, or putrid. Even hyenas get most of their protein by hunting, contrary to popular belief.[8]

The coolest how-did-we-get-meat proposal is the endurance-running hypothesis. Most land mammals reduce heat load by panting, which channels away body heat through secretion of saliva, along with some sweating. When the animal is not in rapid motion, air trapped within its fur insulates it from overheating or from getting too cold. (Large African mammals such as elephants and hippopotamuses are hairless because they have a greater proportion of body mass to surface area and thus are in danger of internally overheating.) Panting and air-trapping fur work brilliantly for brief surges of power interspersed with long stretches of laziness, but champion sprinters like cheetahs have to call it quits after a quick dash because heat buildup could kill them outright.

By contrast, early *Homo* had an upright posture and was likely hairless. Running generates airflow over a naked body, sucking away heat; since mammalian fur traps air, it prevents this air convection effect, especially if the fur is matted down from moisture. In addition, primates are unique in having special eccrine sweat glands, which are scattered over the surface of the skin and dissipate heat through fine perspiration. Running efficiency increases in proportion to leg length, and thus *Homo* evolved longer legs, as well as Achilles tendons to absorb and release spring energy and larger buttock muscles to augment running power. As a result of these adaptations, humans are able to outrun and capture four-legged, fur-insulated animals such as small antelopes, kangaroos, and hares over long distances in high heat conditions. Endurance running could also have enabled early *Homo* to reach fresh carcasses killed by big predators more quickly, before heat and scavengers finished off the dinner for good.[9] However, the endurance-running hypothesis also faces some problems. Endurance running is great for pursuing animals over desert terrain, where hunters can follow tracks in sand for miles, but it's unlikely to be a useful strategy in brush or forest. Between 3 million and 2 million years ago, our ancestors may have lived in grasslands, in forests, or by lakes and rivers, so endurance running probably did not work for many or most of them.

While the debate over how our ancestors obtained meat continues among paleoanthropologists, it's reasonable to conclude that our ancestors were, like their orangutan, gorilla, bonobo, and chimp relations, strongly motivated to get their grubby little hands on a piece of meat. Our ancestors, though, were better than their relations at running down prey, better at lobbing stones at scavengers, better at making tools, better at picking up nifty hunting tricks from their peers and elders. As a result, our ancestors had more and more meat to eat as time went on.

Did getting more meat allow our ancestors to evolve bigger brains? It's true that around 2 million years ago our ancestors started acquiring a lot more meat; their brains evolved to increase dramatically in volume, but eating more meat wasn't necessarily the reason behind the brain boom. Predators like sharks and alligators eat a lot of animal protein but aren't reputed for being especially bright. Conversely, gorillas, orangutans, and elephants eat very little animal protein but are well recognized for their intelligence, so animal protein isn't necessary to evolve intelligence, either.

There is another link between meat and brains: Animals that are good at coordinated team combat against one another are also good at hunting other species. Male chimps, working in teams, can kill other chimps that happen to be caught without allies nearby. They also pursue monkeys in coordinated assaults and are able to acquire around 10 percent of their calories from prey like colobus monkeys (chimps are bigger and smarter than monkeys). Male chimpanzee hunting packs can decimate populations of monkeys despite the agility of the latter. Other animals that are good at working together to hunt big animals, like wolves and hyenas, use their fangs and teamwork to eliminate rivals. Wolves have high kill rates, especially during the winter, when groups of them may encounter a straggler. Conversely, animals that do not team up against one another in combat tend to make poor group hunters. Bonobo apes are close relations of chimps but do not engage in lethal coordinated combat against one another; they also do not hunt in teams and instead individually catch small animal prey when chance occasions arise.[10]

It's possible that our ancestors hunted in coordinated packs, using cunning instead of brawn and fangs. There's every reason to believe that our *Homo* ancestors were more skilled at killing one another and hunting animals than chimps, using weapons like clubs, spears, and rocks, in teamwork with other assailants. Around 2 million years ago, *Homo* skeletons showed signs of adaptation to a terrestrial lifestyle— shorter toes, narrower pelvises, the neck shifting to support the skull in an upright position, the upper thigh bones (femurs) tilting inward to improve gait. Braincase volume more than doubled from about fifteen ounces in Australopithecines to thirty-two ounces in our *Homo erectus* predecessor.[11] Thus the increase in brain size was closely related to the shift to walking fully upright. Walking around on the ground all the time—and especially sleeping on the ground—would have been a hazardous habit for our ancestors in territory inhabited by lions and other predators, unless our ancestors themselves had become formidable creatures.

The primatologist Richard Wrangham argues that the mastery of fire and cooking could have enabled our *Homo* ancestors to obtain more calories from raw meat and starches; fires could also have helped fend off predators at night. It's an intriguing proposal, though for the moment the cooking hypothesis remains unproven, awaiting evidence of fire usage 2 million years ago. Solid evidence for the use of fire in Europe currently dates back to 300,000 to 400,000 years ago.[12] Alternatively, instead of running down prey, perhaps our *Homo* ancestors were using their hands coupled with ingenuity. To learn more about the history of stone tools, I headed to Africa, the place where our ancestors first mastered the art of tool making.

Through the management of a bed and breakfast in Nairobi, Kenya, I arrange for a driver to take me out to Olorgesailie. No one that I talk to has ever heard of the place, including the driver, so I show them the location on Google Maps. It's early afternoon by the time we set out to the site, and the sweltering heat has subsided. The road ascends over a ridge, the dilapidated buildings of the city soon giving way to an expanse of acacia and burnished red sands. The driver swings the car wildly back and forth to avoid potholes. He doesn't speak much, except to ask for directions. We pass Maasai, wearing cloaks or school uniforms, loping with a graceful stride. An old man and a little boy hop into the car to guide us and catch a lift. We finally spot the sign directing us to an open-air museum, Olorgesailie Prehistoric Site, operated by the National Museums of Kenya.

A guide leads me to the first exhibit. Scattered in the sand, beneath a wooden catwalk, are dozens of rocks. Something about their shapes immediately catches the eye. The stones are approximately the length of a hand or two. Their silhouettes are pleasing, resembling almonds or perhaps teardrops shed by a grieving volcano.

Nestled within the arc of the Great Rift Valley, Olorgesailie is one of the most enigmatic sites in our evolutionary history, as inscrutable as Stonehenge, but vaster in geographical scope and implications for the development of humankind. Gathered at this site are hundreds of rocks that were meticulously crafted approximately 800,000 years ago and then left inexplicably behind. These rocks were sculpted in this fashion starting as far back as nearly 2 million years ago, with little variation in design for over a million years.[13] The same design was found in Africa, Europe, and Asia, so the stones must have been indispensable for doing *something*—but what? Most anthropologists surmise that these rocks were Paleolithic Swiss Army knives, capable of carrying out a range of functions, such as cutting meat and scraping hides. These Acheulean hand-axes can indeed be held, in inverted fashion, with the bulge of the tear pointing upward, but because the rocks were sharpened on both sides, they're uncomfortable to grip tightly. Imagine trying to wield a butcher knife that has a cutting edge on both sides and no handle. Other dedicated stone tools found at the same time as these teardrop rocks seem much better suited for the purposes of cutting and scraping.

Another problem with the Swiss Army knife idea is that in some places, large quantities of the tools were made and then apparently abandoned. It's been proposed that the makers may have simply forgotten where they had laid their finished products. Alternatively, these stones may have been cores from which smaller, more useful blades were hacked off. This raises the question of why these leftover cores were shaped in the striking teardrop fashion, with the center of gravity displaced to one side, rather than being molded with the center of gravity dead in the middle, as you might expect from simple raw material for making blades.

One recent conjecture is that the hand-axes were made by men to court women, with onlookers blushing over the teardrops that emerged as the most symmetrical and skillful. As the anthropologist Steven Mithen proposed, "The thrill of holding a finely made symmetrical hand-axe is an echo of the Stone Age past . . . when these objects played a key role in sexual display."[14]

It has also been pointed out that the proportions of many Acheulean hand-axes conform to the golden ratio (the ratio of the length to the width is the same as the ratio of their sum to the length) that the ancient Greeks and others held in such high esteem in their buildings.[15] But could such an object help Great-grandpa get Great-grandma into the bush? Some large hand-axes, on the order of twelve inches long and four and a half pounds or larger, seem poorly suited for practical use, and may have functioned as show-off items, but the idea that hand-axes were made to woo women has attracted scorn from some anthropologists.[16]

Another suggestion is that the smaller hand-axes were chucked rather than wielded. This gets around the problem of the sharp edges—knives are typically thrown from their blades. Since chimpanzees and gorillas can throw sticks and stones, it's arguable that 2 million years ago our *Homo erectus* relation could have been a formidable thrower. In the human record, there are many other examples of weapons that were thrown. The best known are the boomerangs made by Australian and Tasmanian groups. In the hands of a skillful thrower, nonreturning boomerangs could deliver a formidable wound if thrown overhand or knock down large animals if thrown horizontally at their legs. Similar throwing sticks have been discovered in Peru and Africa.[17]

When I learned that some researchers believed hand-axes may have been thrown discus-style and that many others pooh-poohed the hand-axe-throwing idea altogether, I asked Professor Gail Kennedy at the University of California, Los Angeles, if I could borrow a hand-axe to try some experimental throws.[18] To my surprise, the professor handed me a real Acheulean hand-axe from her personal collection. The hand-axe felt heavy when I tossed it with an overarm throw at some branch targets laid on a grass embankment. Just a few throws wore me out, and my hand was gashed. Nonetheless, there was a fascinating, macabre heft to the stone tool.

It's unlikely that a single thrown hand-axe could have killed a mammal. However, prior to the invention of bows and arrows and other projectile weapons, if *Homo erectus* were to make a throwing weapon out of stone, it could either be shaped like a baseball, which could be thrown fast and far but would bounce off the hide or skull of a large mammal, or it could have a cutting edge, in which case something like an Acheulean hand-axe would have been the logical design, the closest a stone tool creator could ever get to working basalt or chert into a decent throwing knife. In all likelihood, it still wouldn't kill, but with skill and luck it could open a wound, and the prey might be pursued afterward. If throwing weapons were used around a body of water, such as existed near Olorgesailie, then a stockpile would have been necessary to replace the lost arsenal. Carrying them in a bag like a sack of quivers, a *Homo erectus* hunter could have had at least a few chances to bring down some game or to inflict a serious wound on an enemy. It's hard for us today to imagine how a walking ape hurtling a sharpened rock could bring down an animal, but perhaps a thousand years from now, it will be equally hard for people to imagine how a baseball could be thrown at a hundred miles an hour into a catcher's mitt sixty feet away. Whichever hand-axe theory anthropologists support, they all agree that the skill and imagination required to make these objects is wondrous to contemplate and helps to demolish the stereotype of a brutish hunter-gatherer shuffling about in the wilds, devoid of creativity and refinement.

One time, while studying anthropology in Los Angeles, I found myself with a few days off. Not having made many friends in L.A. by that point, I packed up a tent, a stove, some tins of octopus and sardines, and a couple of gallons of water in plastic jugs and piled everything into the hatchback of my '92 Ford Escort. During a recent teaching session, an archaeology student had demonstrated the art of knapping, or shaping a stone tool—an Acheulean hand-axe, in this case. The exercise seemed dangerous but fun, so I headed off to the desert to see if I could bang out some hand-axe replicas myself. I set up camp in a valley without a soul in sight. Overlooking my campsite was a long-abandoned dwelling high on top of a hill. I hiked up the boulder slope and explored the foundation ruins, wondering what kind of man would choose to lead his family out here and how long it took before the wife packed up her suitcases and left her misfit ex-husband alone to ponder life in this desolate spot. The landscape was painted in ochre and sandstone, peppered with forlorn Joshua trees, yucca, and low thorny bushes, the shadows distinct and long. I picked up different kinds of rocks and chipped away at them, and eventually, after cutting my hands, succeeded in creating a crude hand-axe, with none of the smooth curves on the Acheulean hand-axes but of approximately the right proportions.

There were a few skinny gray hares hopping about. I picked up my hand-axe and stalked them, trying to ding one with the weapon. I could not even get within a few feet of hitting one because the hares were too fast and wary and my aim was terrible. Hungry, discouraged, and dusty with all the running about, I gave up and started back to camp. Perhaps that night I would crack open a tin of sardines in tomato sauce and gaze at the stars; perhaps, before the sun disappeared

altogether, I would have time to carve my initials into the hand-axe—
or someone else's initials? I could spend the night under the blazing
stars pondering which lady-friend would be most impressed with the
symmetry of my crude handiwork. Who knows, if the fellow up on the
hill had been a better knapper, he might have been able to woo another wife for company in this lonely desert.

The Paleo diet, often slammed by mainstream nutritionists as one of
the worst contemporary diet plans, takes its name from the interval
when stone tools first appear in the archaeological record, more than
2 million years ago. Similar to the Atkins/low-carb diet, it embraces
meat and fat. The standard argument is that humans evolved to eat
meat, fish, vegetables, and occasionally fruit and tubers, and any dietary innovations that came afterward, like milk, wheat, potatoes,
corn, and beans, were too recent for evolution to have caught up and
modified our genes and digestive systems. (Although people who believe in human evolution and those who subscribe to literal interpretations of the Bible have generally quarreled in the past, some of them
now find themselves strange bedfellows in their unity concerning the
virtues of meat.) The Paleo diet looks and sounds evolutionary, but
some paleoanthropologists dismiss it as overly simplistic, a caricature
of actual evidence—like dividing the world into good guys and
bad guys.

Supporters of the Paleo diet argue that it was not just the Inuit who
could thrive on a meat-rich diet. They point to Vilhjalmur Stefansson,
a second-generation Icelandic American anthropologist, writer, and
Arctic explorer in the beginning of the twentieth century, as one example. Stefansson and one of his Danish expedition mates, Karsten
Andersen, thrived on boiled mutton and mutton broth for a year,
thereby settling a long-standing debate about whether it was possible
for humans to live on meat alone.[19] All the meat eating on his Arctic
expeditions didn't seem to adversely affect Stefansson's health. He was
romantically involved with the American novelist Fannie Hurst, had

a child with an Inuit woman, and finally settled down with a twenty-eight-year-old woman when he was sixty-two. He passed away from a stroke at the age of eighty-two.

Recent genetic studies offer potential backing for Paleo and low-carb diets in combating obesity. It turns out that people vary in the number of genes required to produce salivary amylase, an enzyme used to break down starches in the mouth. Most people have around five copies of this gene; the overall range is between two and thirteen. People who have fewer copies of this gene are more likely to be obese. In theory, this means that eating fewer starches should help these people lose weight, but this depends critically on what people eat instead of starches. Substituting fat for starch increases the palatability of food, which could make people eat more; substituting animal protein for starch may trigger similar problems. Substituting plant protein for starch might seem like a good idea, but then too much protein leads to protein intoxication and lack of palatability, which will tempt dieters to binge on foods that taste good, like starches, fat, and meat, and we're back to square one. In the future, it might be possible to take pills that mimic the effect of salivary amylase. In the meantime, as we will discuss later in the book, the only measure so far that is effective in reducing weight is to increase moderate physical activity, principally walking, and to decrease sitting time.[20]

Committing to carb avoidance means calories have to come from somewhere other than starches or protein; humans can only tolerate protein consumption comprising a maximum of about 40 percent of total caloric intake, due to the toxicity of nitrogen compounds stemming from protein digestion. (During his yearlong meat-only experiment, Stefansson insisted that he be allowed to eat meat with a lot of fat on it, to counteract the effects of eating a lot of protein.) This leaves saturated fat as a major source of Paleo energy, because there's only so much olive, avocado, or fish oil a person would want to consume. Rather than starve, some Paleo enthusiasts plow into fatty meat. The problem is the supermarket version of Paleolithic diets: A day's worth of stalking wild game and snacking on bugs is replaced by beef steaks, sausages, pork chops, and fried eggs, fatty rich fare beyond the dreams of most hunter-gatherers. Few if any nutritionists would object to a historical hunter-gatherer diet, whether it be based on deer or nuts and grasses.

Paleo and low-carb enthusiasts, however, are apt to characterize these kinds of criticisms as nitpicking, countering with a legitimate observation: They simply feel and perform better—at the office, in the bedroom, at the gym—eating a lot of meat, fat, and cholesterol; so take that, you lousy politically correct pudgy carb-eaters. As it turns out, there is some scientific justification behind the connection between meat, mood, and sex. One key aspect in this connection is cholesterol. Our livers and intestines synthesize most of our cholesterol, but in Western diets, 12 percent to 15 percent of serum cholesterol comes from dietary sources such as eggs, oysters, whole milk, and meat.[21] Humans use cholesterol in a wide array of body tissues and to manufacture hormones like cortisol, estrogen, and testosterone. Women have far lower levels of testosterone than men, but the hormone is still critical for sex drive in women. Flagging libido can be treated using testosterone patches, creams, or injections, but pundits throughout history have advised using cholesterol-rich foods to spice up your love life. Brains, crustaceans, mollusks, cuttlefish, octopus, and oysters were used as aphrodisiacs in ancient Greece, imperial Rome, and medieval Europe.[22] Oysters were potent symbols of eroticism in seventeenth-century Flemish allegorical paintings.[23] T. S. Eliot made reference to the association between oysters and lust in his gritty poem "The Love Song of J. Alfred Prufrock":

*Of restless nights in one-night cheap hotels*
*And sawdust restaurants with oyster-shells*

Writing in the eleventh century, Constantinus Africanus described the following cholesterol-rich aphrodisiac:

*Another medicine which is taken before intercourse because it is amazingly stimulating: take the brains of thirty male sparrows and steep them for a very long time in a glass pot; take an equal amount of the grease surrounding the kidneys of a freshly-killed billy-goat, dissolve it on the fire, add the brains and as much honey as needed, mix it in the dish and cook until it becomes hard; make it into pills like filberts and give one before intercourse.*[24]

The sexual powers of lobster were acknowledged in a poem written in 1713:

*The Lusty Food helps Female Neighbours,*
*Promotes their Husband's, and their Labours;*
*And in return much Work supplies*
*For that Bright Midwife of the Skies.*
*Lobster with Cavear in fit Places,*
*Gives won'drous Help in barren Cases;*
*It warms the chiller Veins, and proves*
*A kind Incentive to our Loves;*
*It is a Philter, and High Diet,*
*That lets no Lady sleep in Quiet.*[25]

Consumption of cholesterol and fats of all kinds (except trans fats, which are found mostly in industrially produced foods and red meats) also props up high-density lipoprotein (HDL) cholesterol levels.[26] When HDL cholesterol tanks, men are at greater risk of impotence and erectile dysfunction. Nuts, employed as aphrodisiacs by ancient Greeks, also boost HDL cholesterol levels.[27] Widely used cholesterol-lowering statin drugs inadvertently suppress testosterone and increase the risks of erectile dysfunction and diminished libido.[28] People with low cholesterol levels are also more likely to be irritable or depressed, get suspended or expelled from school, and perish from violent deaths, including accidents, homicides, and suicide.[29]

Thus there is considerable scientific evidence that eating generous portions of animal foods is likely to put one in a good mood. On the other hand, eating a lot of meat likely predisposes girls to reach sexual maturity at an earlier age and thereby die sooner as well.[30] By the cold calculus of natural selection, that's an acceptable compromise, because it would have meant more babies starting out life earlier. Evolution doesn't necessarily favor animals that live longer; all else being equal, evolution favors animals that have more compact life spans, reproducing and dying sooner, for the same reason that a nimble company that produces stylish but cheap gadgets or clothing can outcompete brands that take longer to adapt and reach the market.

This is the "life-history" view of diet and health: More robust health at an early age comes at the expense of longevity. Prostate cancer also has the hallmarks of a life-history disease. The nutrients that tend to put men at higher risk of getting prostate cancer—calcium, zinc, fat—were scarce in ancestral diets, but a diet rich in them, along with higher calorie intake in general, would make a man taller, more buff, and richer in sperm count, and thus a stronger contender in the mating market.[31]

In other words, the robustness of meat-eaters and the long lives of meat-abstainers are two sides of the same biological coin. It all depends on how you define *healthy*. Does healthy mean being in a great mood and being fertile and stronger at a younger age, or does healthy mean delaying cancer for a couple of years and hanging out with your great-grandchildren? This is a question that each of us—and especially parents—needs to carefully consider when thinking about the Paleo diet and other meat-heavy regimes.

Ah, there's someone else knocking on the door. . . . Everyone, please make room, and I mean a lot of room, because I'd like to introduce a final guest at history's meat-eating table: your cannibal cousin.

The evidence for cannibalism is omnipresent throughout the animal kingdom. Insects, spiders, leeches, octopus, fish, salamanders, frogs, and birds do it; so do mammals ranging from mice to polar bears, gorillas, and chimpanzees; so did our hominin ancestors, in places like Spain and Iran and China, judging from cut-marks on bones, telltale traces in feces, and cooking residues; and so did modern humans, all around the world. Like most animal cannibals, hominins usually ate infants and juveniles because they put up less of a fight, though fallen enemies were fine dining (or treated like garbage) and deceased relatives were honored.[32] Everything was gobbled, including muscle, marrow, and brain, except perhaps the gallbladder, reported to be bitter.[33] The cannibalistic habit was so widespread that it may have even left a genetic signature in our DNA, a gene variant that confers resistance to a disease from eating prion-infected brains.[34] When viewed in broad perspective, what's most notable about human cannibalism is how squeamish we've become about it.[35] As we'll discuss in the next chapter, in part that's because humans view flesh as more than just food—it's a cultural hydra writhing with taboos and scandal.

# THE PARADOX OF FISH

*Xin dung che mam tanh hoi.*
*Co mam co ruoc moi roi bua an.*
Please don't turn up your nose at smelly fish sauce.
Only with fish sauce and pickled shrimp do you get a
real meal.

*—Vietnamese saying*

My mother never returned to visit Vietnam after she emigrated to Canada, but half of my genes are from her, and so as my plane lands in Saigon, it is as if she has come back to the homeland, the ghost of her DNA expressed in my features—the same dry skin prone to eczema, the same thin hair—and behavioral tendencies, like an aversion to noise, crowds, and being rushed. I've booked a windowless third-story hotel room in gringo-grunge Pham Ngu Lao. The hotel driver who picks me up at the airport is initially taciturn, but when I tell him that I am writing a book about traditional food, he becomes animated in describing his Mekong Delta hometown specialty, *banh xeo*, a kind of pancake that features bean sprouts, assorted greens, and shrimp in a quick-fried rice-paper wrap.

"You say you're Vietnamese and you don't know *banh xeo*?" he exclaims.

It's been five years since I last passed through Saigon, and I struggle to decipher the driver's slippery southern Vietnamese patter, with its emasculated consonants. The streets in Saigon seem more vivid, more fluorescent, denser, busier, cleaner. They sprout at angles and places that I don't recognize.

Wishing to learn about Vietnam's famous fermented fish sauce, I surf the Internet at my hotel and come across an article about a fish sauce entrepreneur, Hang Thi Dao. As bratty boys growing up in Canada, when my brothers and I caught the stink of Vietnamese fermented fish sauce wafting from a pot bubbling on the stove, we would shriek like monkeys and flee to the basement. (Fermented soybean paste, brought out onto the table to season boiled pork and shrimp, was found even more repellant.) But that was more than thirty years ago; a lot can change in thirty years. I look up Hang on Facebook. It turns out that we have a mutual friend, so I send her an introductory message.

A few days later I board an airplane to Da Nang on the southern coast. From there, I catch a bus with Hang's earnest and polite younger brother. I buy him and myself a Vietnamese *banh mi* (submarine sandwich), oily processed meats set in cilantro, butter, pickled radish and carrots, in a paper-crisp baguette that shatters upon biting. The seating on the bus is cramped, and there's no air-conditioning on a humid July morning, but since Hang's brother peppers me with questions about society and economics, the time flies by.

We arrive at Hang's family home, where the low house faces the trucks and long-distance buses hurtling along Highway 1. Hang welcomes me with a broad, smiling, guileless face, as if we've been friends for decades. She proudly shows me a papaya tree in the backyard just starting to bear fruit. A sow grunts from the family pigsty while I wash off grime from the journey. Behind the house are the remnants of a long, crumbling runway, interspersed with patches of tough weeds. Located next to the demilitarized zone, the province of Quang Tri was bombed with the greatest proportion of ordnance during the Vietnam War, much of it still unexploded and a menace to the local populace.

Hang and her brother take me to a river where their father used to work as a fisherman. The fish, however, are mostly gone. Sand dredging destroyed their habitat, Hang explains. To satisfy the demands of construction work, machines were brought down to the river to extract sand. As a result, there were landslides, houses collapsed, and families were forced to migrate. Moreover, the riverbeds here used to be rich habitats for shrimp, mussels, and fish. Hang says that now all that remains in this river is water and sand. To compound difficulties, Central Vietnam is known as the poorest area of the country due to geographical bad luck: The summers are searing, while during the monsoon months, rains flood the land and cause havoc. The eldest of eight children, Hang remembers walking to school while other students rode by on bicycles, unwilling to associate with her because of her family's poverty. "My family didn't have a clock. I just got up when it was dark. Sometimes I arrived at school when no one was there," she recalls. From the age of twelve, she helped her father fish on the river or sold the catch to farmers in the mountains. When farmers lacked cash, she bartered the fish and prawns for cassava, rice, sweet potatoes, and other produce. A hardworking student, Hang earned a scholarship to study agriculture in Hue and then another one for a master's degree in Australia in sustainable development. Inspired by a mentor who pointed out to her the value of Vietnamese traditional cuisine, Hang returned to Quang Tri to set up a new brand of fish sauce. Produced by the farmers and fishermen in her area without the use of artificial chemicals, it is bottled under the brand name of Bamboo Boat (Thuyen Nan), a reference to her humble past.

As the sun sets over the old runway, the sky transforms from vivid blue to searing violet, an intense pureness of color rarely witnessed in smog-wreathed East Asia. When we return from the river, Hang, her mother, and three of her brothers sit down to eat around a table set up in front of the house and the highway traffic rushing by. Arrayed about the table are the elements of a Central Vietnamese rural feast: fried fish, caramelized pork, squash soup with minnows, spicy pickled prawns, two kinds of pickled fish, a dish of tart greens, cucumbers, rice noodles, and the most extraordinary fish sauce I have had the pleasure of dipping my chopsticks into: thick, almost creamy, oozing with velvety flavors.

When the meal is over, the children hover at the edge of the highway, watching the trucks and buses, calling out to friends and neighbors. I settle down on one half of a wooden bed, Hang's oldest brother climbs onto the other side, and the mosquito net is secured for the night.

Fish and other small sea creatures comprise the lifeblood of coastal Vietnam. However, catches have been dwindling over the past few decades, and the Vietnamese are resorting to eating smaller fish. This means that more fishermen may end up relying on making fermented fish sauce to support their families. In the short term, this may give fishermen's families an alternative source of income—and give a boost to fermented fish sauce businesses like Hang's—but in the long term, the intensified pressure on populations of smaller fish may prove to be unsustainable. It's a disturbing scenario, especially for an already-poor country like Vietnam. Meanwhile, at a 2013 Tokyo fish market auction, a 490-pound bluefin tuna reeled in a $1.7 million bid; this translates into roughly $250 for a single one-ounce sushi serving of the behemoth's flesh. Tuna were once considered garbage fish by fishermen on the northeast coast of the United States, but now fish flesh has become the newest star on the nutritional stage, acclaimed for its stores of miraculous omega-3 fatty acids and vitamin D.[1]

High-quality sushi is an orgasmic experience in its own right, but paradoxically, quite a few people would be happy to pass on a bite. As scholars have documented, taboos against eating fish were once observed among groups in Afghanistan, Pakistan, India, Central Asia, Tibet, Mongolia, northern Thailand, many regions in Africa, England and Belgium in the Iron Age, Tasmania, and Fiji, as well as among the Norse in Greenland and North American Indian tribes like the Zuni, Hopi, Navajo, Apache, Crow, Kiowas, Comanche, and Niitsi-tapi (Blackfoot).[2] To all these traditional fish haters, we could add a few children and adults today. In my house, when we ate a lot of fish, my mother would sometimes mutter, "I'm eating so much fish, I'm becoming one!"

When asked why they avoided fish, people gave many answers: Fish looked like snakes; fish ate people's corpses and therefore eating fish would be equivalent to an act of cannibalism; water was sacred and therefore fish were sacred; fish were unclean; fish could not cry for help or mercy, so killing them was especially cruel; eating fish would cause one's teeth to fall out; most commonly, they said eating fish was simply disgusting. All of these explanations may have been quite real to the noneaters, but they still do not answer the question of why so many different people around the world felt (and feel) dire revulsion at the thought of dining on a source of animal protein and fat that happened to have fins instead of feet. To make matters more complicated, many people who avoid eating meat are often fine with eating fish.[3]

The first drawback to eating fish that comes to mind may be the bones. Ingestion of fish bones carries the risk of piercing the esophagus or intestines, and triangular fish bones, such as those located around fish heads, can be tricky to extract from the esophagus. A second drawback is that fish meat is generally lean, and while that seems fine considering our current fat-abundant diets, too much protein in a diet can be an issue. Another concern is that top-of-the-food-chain tropical fish may accumulate toxins from a marine plankton (*Gambierdiscus toxicus*), which can cause ciguatera poisoning. Its symptoms—including nausea, intense vomiting, diarrhea, and paralysis may persist for several years, and it can lead to coma and even death. Worldwide, between ten thousand and fifty thousand people are poisoned by ciguatera toxins annually. Carnivorous fish may also accumulate toxins from feeding on plants, worms, mollusks, corals, and other toxic fish.[4] In recent decades, larger fish have also been noted for their tendencies to accumulate mercury, PCBs, and other toxins from human-made pollution in their flesh.

Ironically, other disadvantages of eating fish in traditional times may have stemmed from the very reasons that fish are now celebrated: omega-3 fatty acids and vitamin D. Cold-water deep-sea fish have bodies that are replete with omega-3 fatty acids, since the structural flexibility of these fatty acids allows the fish's body to compress and expand in response to changes in depth and pressure and maintain membrane fluidity in a cold environment.[5] Humans cannot synthesize omega-3 or omega-6 fatty acids from scratch.[6] If either is completely removed from the diets of children, growth is impaired. However, despite the fact that they are both essential, omega-3 and omega-6 fatty acids often have opposing functions in the human body, with omega-3 generally decreasing inflammatory reactions (the sequence of pain, swelling, heat, and healing of wounds and infections) and omega-6 generally increasing inflammation.[7]

A diet of wild or traditionally raised foods has a ratio of omega-3 to omega-6 of around 1:1, but over time, that balance has skewed toward a greater proportion of omega-6 fatty acids, particularly in industrialized countries, where omega-6 is common in cooking oils and processed foods. The dietary ratio of omega-3 to omega-6 has been estimated at 1:2 among rural South Asian Indians, 1:4 among the general Japanese population, 1:6 among urban Indians (South Asia), 1:8 among Australians and Belgians, 1:9 among twenty-something Japanese, and 1:10 among Americans. In 1909, omega-6-heavy vegetable oils combined did not make up even half a percent of calories consumed in the USA, but by 1999, they provided nearly 10 percent of all calories consumed, with soybean oil alone accounting for 7 percent. The major impetus for the newfound devotion to vegetable oils in the American diet was the decision by politicians and health authorities to shift their efforts to eliminating saturated fats, beginning in the late 1960s, as part of the assault on heart disease. Feeding livestock omega-6-rich seeds instead of grasses and insects and the longer shelf life of omega-6 fatty acids also helped swing the pendulum in favor of omega-6 fatty acids in Western diets. Diets replete with these rough-neck omega-6 Rambos have been investigated for possibly delaying recovery from surgery and trauma and exacerbating autoimmune diseases, heart diseases, obesity, depression, and bipolar disorder.[8]

On the negative side, high serum levels of omega-3 fatty acids have been associated with more aggressive prostate cancer. More worrisome for people in preindustrial societies, however, would have been the tendency for omega-3 fatty acids to increase bleeding incidents and bleeding time (omega-3 fatty acids are runny), a problem that the Inuit had to contend with.[9]

Besides omega-3 fatty acids, the other reason fish are heralded as the new saviors of health food is that they tend to contain a lot of vitamin D from eating vitamin-D-replete plankton and algae. More vitamin D in the diet might seem like a good thing, but in traditional societies where people worked outside all day, they had all the vitamin D their bodies needed, and taking in more vitamin D could have led to vitamin D intoxication.[10] The Pacific Coast Indians ate a lot of salmon, but archaeologists observe that their children did not eat as much salmon as the adults, perhaps to avoid the effects of vitamin D poisoning, which include kidney stones, nausea, vomiting, headaches, constipation, and elevated levels of calcium in the blood.[11]

Troublesome bones, overly lean flesh, and overdoses of marine toxins, vitamin D, and omega-3 fatty acids can explain some of the aversion that people have often demonstrated toward fish. But we need a theory of food taboos that can explain why fish become acceptable fare in some areas and not in others—fish avoidance was historically common among Bantu speakers in East and South Africa, but neighboring groups such as Bushmen and Hottentots were not necessarily put off by fish—and why other foods, such as meat, milk, and insects, are avoided by some and relished by others.[12]

To solve this puzzle, consider the following problem: If you need to buy a shirt, which color should you select? There are two easy shortcuts to this problem: Buy the color that most people are currently wearing (the follow-the-crowd rule) or buy the color preferred by your favorite athlete or musician or any other public figure (the copy-your-idol rule). Either way, you leverage hidden information that your peers or idols may have about the cool thing to wear, so you won't look like a dork on the street, as many academics tend to do. (An ex-girlfriend once asked me, "Do you dress in the dark?")

Employing heuristics means you won't waste time agonizing over making the right decision in those soul-draining wastelands known as shopping malls. The anthropologists Peter Richerson and Robert Boyd argue that these kinds of quick-and-dirty cognitive shortcuts allow us to acquire information efficiently, but with the result that we sometimes end up acquiring information of dubious value. As they put it, the human capacity to acquire culture was built for speed, not for comfort.[13]

Food taboos challenge us with the same kinds of conundrums. What should we eat? Most animals don't need to worry about this problem, because the information is more or less wired into their brains from birth and constrained by predictable environments. Humans, on the other hand, do worry, because our brains do not come prewired with food preferences. Instead, we're equipped with a raft of heuristics that can be applied to the problem of deciding what to eat and what to reject. Almost choked on a bone at the age of three? Does the food in question smell like sweaty socks? Parents and older siblings like it? Someone you admire and respect likes it? The great advantage of heuristic-based food preferences is that a kid can grow up anywhere in the world and quickly acquire an effective repertoire of safe foods. The principal drawback is that we sometimes end up rejecting perfectly good grub, as every parent knows and fears.

During a family dinner in Sapporo, I quizzed one of my Japanese friends about the ethics of eating whale and dolphin meat. The doctor, normally extremely congenial, but with a few glasses of wine in his system, turned beet red. "People in America eat cows and pigs! What's the difference?" he sputtered. I described the viewpoints of Westerners with regard to dolphins and whales (smart animals, TV shows) versus cows and pigs (long association with barnyard status) but the doctor became ever more irritated, so I decided to drop the matter, lest a perfectly good evening and excellent meal be spoiled. Sometimes our food heuristics lead us into strange dilemmas. To a pescatarian, a fish has membership in the edible category, while mammals belong to the realm of friends; laudable philosophy to some, laughable to others. To the Bantu in East and South Africa, fish were despicable snakelike monsters. To Tibetans, fish were helpless beings that lacked the ability to cry out in pain and therefore deserved compassion.

Just as notions about friendship can be recruited into food psychology, so can cultural rules about eating be drawn into ethnic politics and used to fence off outsiders. When I was growing up in Ottawa,

Canada, English-speaking kids would slur French Canadians with the epithet "frog," apparently referencing their habit of including frog legs in their dietary preferences. I, too, quivered with revulsion at the thought of frog legs passing my lips, but when I overcame my prejudice as a teenager traveling in Quebec, I found the slight flesh of *cuisses de grenouille*, fried in a batter, to be exquisitely tender and savory, better than any chicken wing.

According to the Roman historian Plutarch, an elephant-nosed fish species was worshipped by the Egyptian city of Oxyrhynchus. When the residents of Oxyrhynchus discovered that the residents of Cynopolis were eating the elephant-nosed fish, they retaliated by eating dog, sacred to the Cynopolitans, thereby triggering a civil war.[14] If it seems far-fetched that humans would wage war over someone dining on their city's mascot, consider that in Vietnam, where dog meat is commonly eaten (particularly in northern and north-central Vietnam), dog-kidnappers have been caught and killed by village vigilantes in recent years. Hang, who lost two of her dogs to thieves, told me that she would have happily joined in to thrash the miscreants. (One theory is that dogs may have been originally domesticated as a source of meat.[15]) A Vietnamese ex-soldier recounted to me that when he was interned in a refugee camp on a Malaysian island during the 1970s, refugees caught cooking pork in their dwellings were caned by local authorities. In southwestern Ethiopia, the Walamo were said to be so offended by people eating fowl that they killed such transgressors, though ritual experts were exempted. Clearly, one person's totem is another person's meal.[16]

Fresh fish, generally odorless, quickly decomposes at or above room temperature, exuding its distinctive odor. People in Southeast Asia, as well as ancient Rome, discovered that the rapid decomposition of fish could be controlled and transformed into a tasty and pungent condiment. The technique is ingenious. Small fish like anchovies, sardines, and mackerel are placed in a vat and covered evenly in salt, which draws out the water from the fish and prevents the fish fats from going rancid. Spices, sugar, or rice bran may then be added, and in the case of the Romans, wine as well. The fish flesh, slowly dissolved by enzymes from the fish stomachs, feeds fermentative bacteria. Weights are used to push the fish below the surface of the accumulating liquid; if the fish are exposed to the air, they soon rot. After one year of fermenting in the sun, the amber-colored fish sauce is ready to be drawn off.[17]

Lower-quality fermented fish sauce stinks because it contains too much bacteria, which leads to spoilage. Factory-produced fish sauce contains sugar and added chemicals to boost the flavor of a cheap, watered-down product. Despite growing up with Vietnamese food and traveling all over Vietnam, I never knew what genuine high-quality *nuoc mam* tasted like until Hang e-mailed me an address on the outskirts of Saigon, where her fish sauce, still in the infant phase of production, could be purchased every Sunday. I arrived by motorbike at a house with no signage, no placard, not even any indication of fish sauce bottles. A young man and woman came to the gate. "Hello! I'm a friend of Hang. Is this the shop that sells *nuoc mam*?" I asked cheerfully.

The woman and man ushered me in. A small collection of *nuoc mam* bottles occupied a corner of the room. Inside the house, another young man joined us. The trio conferred among themselves in the lilting Central Vietnamese dialect, vanished into a kitchen area, and soon came out bearing a circular tin tray arrayed with rice noodles, broad leaves of lettuce, thin slices of boiled pork, a bowl of pickled fish, and two bowls of fish sauce, one with added chili. We sat on a reed mat on the floor. This fish sauce was distinctly darker than standard mass-produced factory *nuoc mam*. I wrapped some rice noodles and pork in a leaf of lettuce and dipped it into the fish sauce. The first bite sent a jolt through me—a remarkable medley of salty, velvety flavors, as if the little fish had been transformed into fine whiskey.

With such lively flavors jostling around the tongue, one can understand why fish sauce anchors the cuisine of the Southeast Asia archipelago, from Thailand to the Philippines, and why the Romans also celebrated fish sauce. *Garum* was called for in more than 75 percent

of the dishes listed in the cookbook of the first-century Roman gour-
mand Apicius and was transported in terra-cotta amphoras across the
breadth of the Roman Empire. One *garum* trade route started from
Spain and went east to Lebanon, via Sardinia and Rome; another
route plied the Rhine and Rhone rivers into the heart of Europe, and
across the English Channel to consumers in London and York.[18] The
Roman poet Marcus Valerius Martialis observed, in an epigram on
oysters: "A shellfish, I have just arrived. . . . Now in my extravagance
I thirst for noble garum."[19]

Hang's name and mission are steadily rising in public profile,
thanks to Vietnamese media. Hang is constantly texting, making calls,
checking the Internet, networking, traveling around the country—all
of this from a woman not yet thirty. It's a rare and inspiring combination: a condiment that tastes heavenly, the key to a savory, low-meat,
affordable traditional cuisine; shepherded to market by a visionary
from a humble background; each drop of fish sauce churned and
wrung by families from one of the poorest regions of a developing
country, using small fish in a fermentative process that can take up to
a year to complete. It's like fair-trade coffee but with an odor evocative of boatyards and tidal pools. This area in Central Vietnam is
cursed with agriculturally unproductive sandy soil, so the fish sauce
venture could be an important source of income for locals. Hang is
also committed to raising money for children in the area who are believed to be suffering from the lingering effects of Agent Orange.
Many have serious medical conditions and lack quality care, leaving
them and their families suffering in squalid conditions.

While Vietnamese cuisine, particularly in the central and southern regions, hinges on *nuoc mam*, I grew up with food that was a mixture of Canadian and Vietnamese, potatoes alongside rice, butter next
to fermented fish sauce. Since my brothers and I whined about the
smell of fish sauce—part of the confusion of being a second-generation
immigrant—it was only used sparingly at our meals. To see whether
Hang's traditional *nuoc mam* brand tastes as extraordinary as I believe
it does or whether it's just bias from my knowledge of Hang's methods
and ideals, I purchase a small bottle of dark fish sauce. It's the link to

a culture that I once struggled to accept as my own, and it's important that I understand my ancestral cuisine inside out, beginning with fermented fish sauce. To do that, though, I need to consult some genuine experts.

It's been ten years since I've seen my first cousin Chi (Elder Sister) Vinh and her family. I first met them during my inaugural visit to Vietnam in my twenties, and I knew that Chi Vinh's son, Duc, was a perfectionist over a soup pot, and everyone in the family had strong opinions about good and bad Vietnamese food. They would be perfect to judge the quality of Hang's fish sauce.

Chi Vinh and her engineer husband, Anh Quy, have retired but still live in the same house, with the doorbell that I first rang fifteen years ago. When she sees me, Chi Vinh exclaims, "You are very thin!"

While affluent urban areas in Vietnam like Saigon have been swept up in a tidal wave of obesity in the past two decades due primarily to the replacement of walking and cycling with motorbikes, cars, TV, and video games, my weight has remained more or less the same, creating the illusion that I've lost weight. I reassure Chi Vinh I haven't, not much.

"I just thought you were sick," she responds.

By Vietnamese standards, I'm doing everything wrong: no wife, no children, no stable job, and I have no comfortable fat around my waist. These are not people to mince their words. I pull out Hang's fermented fish sauce.

"A small present for you, Anh Quy and Chi Vinh. It's also part of my research."

My cousins' suspicions that I've gone stark mad seem to deepen. Anh Quy, ever the analyst, takes off his glasses and turns the bottle around and around, examining the dark liquid. Chi Vinh asks, "Where was it produced?"

I tell them it's from Quang Tri, a poor province in Central Vietnam. Chi Vinh really thinks I've lost it. "Quang Tri! What does Quang Tri have to do with fish sauce?? Phan Thiet and Phu Quoc are famous for fish sauce."

The dinner table is piled with lovely Vietnamese dishes: papaya

salad with pork, shrimp, and peanuts drenched in a vinegary dressing; and a big bowl of sour fish soup with pineapple, the family favorite. While five kids wolf down the feast, I remind My-Hanh, Chi Vinh's
daughter, about Hang's fish sauce, and she brings it out to the table.
My request seems a trifle absurd, this introduction of a Vietnamese
food product among a family of food connoisseurs who were born and
raised in the country. Everyone stares at the small bottle. The label
has no fancy lettering or flashy colors. My-Hanh pours out a little dish
of the dark amber liquid. They dip their fish into the sauce. I pray
that no one gets an upset stomach. Suddenly, My-Hanh's husband
blurts out: "This fish sauce is very good!"

I'm amazed. My-Hanh's husband and I have exchanged perhaps
ten words in fifteen years. He seems to regard me as akin to lint, a
minor inconvenience that came with the rest of his wife's belongings.
Yet here he is, eagerly dipping bits of fish into musky Bamboo Boat
fish sauce, his face lit up as if he's just encountered a long-lost friend.
The rest of the family also begin to show signs of delight at the earthy
flavors of Hang's fish sauce. My-Hanh asks me where she can buy
more Bamboo Boat fish sauce and how much it costs. The price is a
significant premium over the regular factory-made fish sauce that they
buy, but I sense that Bamboo Boat will pull in the customers nonetheless, once Vietnamese are reawakened to the pleasures of traditional
handmade food that harkens back to their ancestors.

*Nuoc mam* is a condiment of coastal Southeast Asia (and formerly ancient Rome), but in mountainous areas of Vietnam, it's considered a
luxury. Family matriarch Aunt Tam told me in Ottawa that when she
was a girl growing up in northern Vietnam, only the rich families
could afford fish sauce. Everyone else resorted to *tuong dau,* fermented
soybean paste. I asked her for the recipe for *tuong dau,* but she told me
it was too complicated to write down. She brought a bottle of her
home-brewed fermented soybean sauce to our house. It smelled like
old shoes and tasted like tofu would if it went to a bar, got drunk, was
mugged on the way home, and woke up with a hangover. Nonetheless,

it's a famous seasoning for committed Vietnamese vegans who abstain from fish sauce.

When I fly to northern Vietnam to learn more about fermented soybean sauce, everyone tells me the same thing: go to Hung Yen, famous for the production of this condiment. I call up my old friend Ly to help me with translation and also as an excuse to catch up. Ly and I know each other from days of dancing Argentine tango in Hanoi. Early on a Saturday morning, Ly is still as feisty as ever. I know I'm in the company of foodies because it takes us an interminable drive through Hanoi and the outskirts to find a place to eat breakfast. I glance longingly at all the noodle and bread stands through the van window, but Ly and the driver dismiss successive eating options that seem perfectly acceptable, all packed with diners at seven in the morning. We continue along several miles of asphalt until Ly and the driver agree on a grungy restaurant at the edge of the highway. We dig into beef noodle soup redolent with fatty flavor and served with a stack of fresh herbs, and accompanied by a special side of tangy bitter melon for me—worth the wait.

After breakfast punches all our pleasure centers, we locate Ban Village. Dozens of shops display racks of fermented soybean sauce by the road. *Tuong ban* fermented soybean sauce, formerly associated with the miserable poverty of Ban Village, has become a signature food item for tourists shuttling along the Hanoi–Ha Long Bay tourist trail. We drop in to visit Thuy, whose name had been independently passed to me twice by friends helping to connect me with the fermented soybean business in Ban Village. Short, smooth-skinned, burdened with a preoccupied air, Thuy agrees to show us around the premises, with one stipulation: "No picture-taking allowed."

I nod my head meekly, like a novice ready to be initiated into a cult. Trieu Son is one of the biggest *tuong* works in Ban Village, bottling five hundred gallons of *tuong ban* each day. Workers carefully step along the rims of stone vats to stir wretched-smelling murky brown liquids. To the left of the stone vat area, molds are reared in a dark room on circular trays of glutinous rice as broad as an arm-span, then mixed in with dried, aged soybeans, along with precise proportions of salt and water. The vats are left to ferment for three months in the winter and half that time in the summer, after which a good batch of *tuong* develops the yellowy-brown sheen of "cockroach wings," as Thuy describes it. Expensive *tuong* is fashioned from mold that was cultured on special fragrant varieties of glutinous rice.

At Dung (pronounced Zung) Nhat, a much smaller *tuong* house in the same neighborhood, the outgoing proprietor, Nguyen Dinh Lap, tells us that the stoneware vats are also critical to the quality of *tuong*. Vats from the province of Ninh Binh, to the south of Hanoi, have the proper mix of earth to produce the best *tuong*.

All morning we've been dipping our fingers into *tuong* vats to sample the fermentation process, but now it's time for a proper *tuong*-based meal. At a house in Ban Village, Ly, the van driver, and I sit cross-legged on the floor with the family around plates of deep-fried tofu, fried pork sausage, sautéed morning glory, pickled spherical eggplant, and squash soup. The interior of the tofu is milky white, fresh and flavorful, with the consistency of Jell-O, set off perfectly by its honey-hued deep-fried skin. When dipped into musky *tuong* jacked up with chilies, it's like a tango between an angel and the Devil, quivering white innocence wrapped in a lustful embrace. Uncle Hai, the van driver, gleefully recites a ditty about the powers of *tuong*:

> *Tuong with medium-rare goat*
> *Eat a piece and you're horny as a goat*
> *Little darling stay here, don't go home*
> *We'll have goat with tuong tomorrow.*[20]

That evening, I meet Ly again to go dancing. The tango hall has the ambience of a decommissioned airplane hanger, the women dressed to kill, the men pacing the sidelines like hyenas, preening and hungry. As Ly and I spin on the floor, she is vivacious as ever, but the day of sampling decomposing soybeans like a connoisseur has taken its toll on me.

"Ly, did you get a stomachache from today?" I ask her.

"No. Why?"

Savage nips at my intestines threaten to topple me on the dance floor. Fermented soybeans may have been the condiment of my ancestors, but the bacteria in my stomach are a meek and callow lot, born and raised in a hypersterile foreign land. I am grateful that the carnage of my lunch stays in my stomach until I stagger up the four flights of stairs to my hotel room.

From the viewpoint of Western nutritional dogma, fermented fish and soybean sauce potentially harbor alarming levels of biogenic amines—microbially produced compounds associated with headaches, rashes, palpitations, hypertension, and diarrhea—as well as high levels of sodium. High sodium levels are also present in Japanese pickled vegetables (*tsukemono*), Korean fermented cabbage, sauerkraut, and fermented cheeses. Sodium inhibits the growth of bacteria that would otherwise spoil the fermenting food. With just enough salt, the bacteria thrive and transform their homes into praiseworthy repasts. Salt is like discipline in a classroom: too much, and the bacteria lose their self-initiative; too little, and all hell breaks loose, to the benefit of no one except the troublemakers.

Curiously, eating *tsukemono* and fermented full-fat dairy have both been associated with longer life spans.[21] Fermented fish sauce also imparts amino acids that are potentially deficient in meat-scarce Southeast Asian diets. Fermented foods like *nuoc mam,* soy sauce, and soybean paste deliver a whammy of umami, or savoriness, transforming a meat-scanty meal into a satisfying feast.

If the umami flavor in this type of diet were reduced, people might compensate for the blandness of their meals by eating more sweet foods or more meat. Eating more meat in turn places serious environmental pressure on land that is far more densely populated than in North America and other Western countries. Fermented flavors help stretch limited ingredients in an area that can't afford meat or fat.

The secret ingredient in *nuoc mam* and other umami foods is the amino acid glutamate, which Japanese scientists have identified as responsible for triggering the sensation of umami, a fifth unique taste, along with bitter, salty, sweet, and sour. Umami is often described as savory or cheesy. Fish sauce is one of the foods densest in glutamate.

Parmesan cheese and marmite are also very high in glutamate. Other foods that harbor glutamate include tomatoes (especially when ripe), potatoes, Chinese cabbage, soybeans, prawns, and Japanese soup stock made from kelp. Mushrooms contain guanylate, which also elicits umami-ness.

However, umami substances alone do not trigger the magic of umami flavor—for that to happen, they must be paired with a nucleotide (the building blocks of DNA) such as inosinate, found in animal flesh like beef, pork, chicken, and fish, which explains why meat is traditionally cooked with glutamate-containing foods like potatoes, tomatoes, mushrooms, milk, cheese, Chinese cabbage, or fish sauce.[22] Glutamate has a long history in human cuisine. However, the infamous monosodium glutamate, better known as MSG, is a different matter.

In 1907, a Japanese chemistry professor, Kikunae Ikeda, discovered how to mass-produce umami flavor by treating wheat gluten with hydrochloric acid, a process that was effective but potentially also dangerous to workers, due to the formation of hydrochloric acid vapors. Ikeda and an entrepreneur founded Ajinomoto, a giant in the food flavoring industry, and the use of MSG in cooking rapidly expanded. MSG, now produced by fermenting sugars, has been consumed for roughly a hundred years, about as long as other mass-produced convenience foods like vegetable oils, white rice, and pasteurized milk. The glutamates that are naturally found in foods are mostly bundled into proteins, like prisoners in a cellblock, and must be digested and released by enzymes before they can exert any effect. MSG, by contrast, is unbound and therefore has potentially stronger physiological effects. Since the topic of "Chinese restaurant syndrome"—a cluster of symptoms including numbness at the back of the neck, general weakness, and palpitations—was first discussed in the *New England Journal of Medicine* in 1968, MSG has been the focus of intense debate among both scientists and consumers. Food giants have fought hard to keep the reputation of MSG from being tarnished by paying researchers to support their product. This tactic has been extraordinarily successful, in view of the regularity with which scientists and mainstream media dismiss MSG concerns as uninformed public hysteria or even racism against Asians.

Nevertheless, bad news still trickles out. Most recently, German researchers have demonstrated that MSG can cause headaches when ingested in high quantities, which is physiologically plausible, given that glutamate is a neurotransmitter and elicits intense pain when injected into muscles.[23] Chinese and Thai researchers have also discovered that higher intakes of MSG are associated with weight gain; not surprising, given that the role of MSG is to make food taste better.[24] Since MSG is ubiquitous in processed foods, under a variety of pseudonyms (for example, autolyzed yeast, sodium caseinate, hydrolyzed vegetable protein), its effect on the epidemic of obesity may be considerable.[25] Concerns about brain damage caused by MSG first surfaced during the late 1960s in tests on mice, but such effects have not been conclusively demonstrated in humans.[26]

One tantalizing question remains: Why do humans enjoy the taste of umami? The four other basic tastes have solid evolutionary credentials: Bitterness helps us avoid being poisoned; sourness helps us avoid foods that are too acidic, such as spoiled food or unripe fruits; sweetness makes us favor high-energy snacks; and saltiness directs us toward sodium, essential in the ancestral environment.[27] Some researchers have speculated that umami makes us attracted to meat and other animal foods. However, raw meat and fish do not have an umami flavor—they're almost flavorless—which may help to explain why people traditionally preferred to cook them with mushrooms, tomatoes, garlic, onions, cheese, and other umamish substances, or else to cook them over a fire, which produces delicious browning of the amino acids (also known as the Maillard reaction, after the French chemist who studied the interaction between amino acids and sugars).

We are still left, however, with the question of why evolution favored umami attraction. It's true that garlic and onions have antibacterial properties, but these are the exceptions, as most umami-containing foods do not. However, umami foods tend to have high concentrations of the amino acid purine. As we discussed earlier, consumption of pu-

rine increases uric acid, which could have been beneficial to our ancestors after they lost the ability to manufacture vitamin C around
60 million years ago. Thus an attraction to umami could be our evolutionary adaptation to acquire antioxidant uric acid. However, these
days, the attraction to the taste of umami is bad news for people who
suffer from the inflammatory condition of gout, because the purine-
rich foods that aggravate their gout attacks, like steak, lobster, and
beer, are plentiful in industrialized countries.

Hang and I catch a bus to Hue, as she attempts to get her fish sauce
certified by visiting an office and meeting some officials. Hang hands
over a wad of cash; she seems optimistic. To celebrate, we head to the
river and relax on the banks. The madness of big Vietnamese cities
like Saigon and Hanoi hasn't come to Hue—yet. We watch the barges
float along the water under a pale blue sky, couples nestled together
on motorbikes. Hang buys sugarcane juice in plastic bags from a side-
walk vendor. I lay out a sheet on the patchy ground, sip the juice,
let the sugar gently rot my teeth as ants march in search of the
sweetness.

Hang spells out her vision of green development for Vietnam:
Rather than the eco-friendly but costly practices of industrialized
countries, she sees traditional rural enterprises like fish sauce as being the way forward. Vietnam is too poor to afford the expensive
eco-friendly practices of Western countries, and traditional rural
enterprises can help develop the country's economy in a sustainable
manner. It's a smart, sensible idea. She tells me about one of her men-
tors, a Vietnamese who married a Swedish woman and came back to
Vietnam to help in the country's reconstruction. He gave Hang a
scholarship to study English at a critical stage, when she had finished
her agricultural studies and needed a high English score to win a foreign scholarship. In Hang's eyes, her mentor represents the ultimate
human being: a person who relinquished the easy pleasures of the industrialized world to nourish the hopes of a desperate nation. Hang

had a chance to stay on in Australia for doctoral studies, but instead she came back to Vietnam, to pursue a life filled with meaning. I ask her if she misses Australia. "Yes, very much," she replies.

If we take a step back and consider fish in the long view of the human diet, they were likely undesirable. Because of their abundance of fine bones, lack of fat, and perhaps preponderance of omega-3 fatty acids and vitamin D, which our ancestors already had plenty of from their daily diet and lifestyles, lean fish would have been far less popular than meat, though fatty and oily fish like salmon and candlefish could certainly be a substitute for land-based animal fare. However, once populations began to settle and meat became scarce, making use of the products of fish and soybean spoilage was a clever way of supplementing shortages of amino acids and boosting the flavor of meatless meals.

As for fish itself, it has undergone a strange transformation in Western nutrition, elevated from woeful pickings to contemporary superfood. However, the rush to fillet the world's remaining stocks of big fish is obviously an unsustainable venture. Nor is it necessary to eat fish for health, because vitamin D can be obtained from adequate exposure to the sun, and omega-3/omega-6 imbalances can be corrected by using less cooking oil, eating fewer processed foods, and shifting to more sustainable protein sources like smaller fish, locally adapted mammals, and insects.

# THE EMPIRE OF STARCHES

Many people do not like to eat vegetables—and the feeling is mutual.

—ADAM DREWNOWSKI AND CARMEN GOMEZ-CARNEROS,
*"Bitter Taste, Phytonutrients, and the Consumer: A Review"*

Most people in Western countries today think of vegetables as healthy. However, as we have seen, cherished Western notions about food and nutrition often turn out to be wrong when viewed in a broad context. Despite the great efforts of Western nutritional scientists to show otherwise, eating plants has never been conclusively shown to improve a person's health prospects; by comparison, drinking moderate amounts of alcohol, consuming moderate amounts of salt, or being somewhat overweight have shown more tangible benefits for overall health. Though the poor results of vegetables and the relatively clearer merits of alcohol, salt, and chubbiness might surprise people in Western societies today, these would have come as no surprise to the vast majority of people in traditional societies. Indeed, they would have been amused or shocked to see how "educated" Westerners today worship salad bars as the quintessence of healthy eating.

The English naturalist Charles Darwin was no poet, but he viewed the natural world with a clarity that was extraordinary for his time, and he captured the dilemma of eating vegetables exquisitely in these (for him, impassioned) lines: "what war . . . between insects, snails, and other animals with birds and beasts of prey—all striving to increase, and all feeding on each other or on the trees or their seeds and seedlings . . . !" In other words, eating plant foods is an act of war, every head of broccoli laid out on the cutting board a decapitation. Our crops are slaves to our hunger; farmer's fields are prisons for thousands and millions of speechless, immobile inmates. I'm not saying this to turn your seven-year-old daughter off veggies forever; I'm saying this because it clarifies why George H. W. Bush and many other people don't like many vegetables, including broccoli. The humorist Roy Blount Jr. conveyed the sentiment in a memorable couplet: "The local groceries are all out of broccoli / Loccoli."[1] Plants may be immobile, but they're far from defenseless.

It's true that many plants have medicinal properties, but this does not make them healthy everyday fare. If plants comprised the bulk of traditional diets, this was partly out of sheer necessity—many large mammals went extinct, and the remaining large animals were difficult to catch, time-consuming to raise, or expensive to purchase—and partly because people learned how to cook vegetables in a way that created a satisfying meal, which is to say, in a way that neutralized the most valiant defenses that a plant could muster against a predator like ourselves.

The goal of this chapter is to analyze the complex history between plants and humans. We'll parse common plants into various categories based on the kind of harm or benefit they confer to us and discuss how humans have evolved, biologically and culturally, to eat plant foods. The overall message: Like most other foods, plants have no nutritional significance on their own; what matters is the overall composition of the meal, the way the food is prepared and cooked, the environmental context, and the genetic ancestry of the eaters.

China is a worthwhile place to consider the question of plant foods, because thousands of years of rice cultivation and high population density mean that animal foods there are relatively scarce and plants foods are paramount.

———

I apply for an English-teaching job in China through an employment agency and receive the green light. Now I'm looking out through dust-caked windows at a gray sky. The college lies at the easternmost edge of a low-slung manufacturing city, Bengbu, formerly renowned as a center for pearl production. Out here at the college, dusty pavement gives way to dirt roads. Clouds of grime billow from trucks ferrying open loads of dirt and gravel. The rear of the school property is demarcated by an oily black waterway, from which water is drawn to feed a patchwork of rice fields. A stand of skinny trees shivers next to the construction road, surrounded by a pile of discarded fertilizer and pesticide bottles and plastic packages. Behind the teachers' apartments, elderly folks hoe, water, and weed plots of cabbage, beans, corn, rape-seed. On the other side of the black canal, tractors grind over the rice plots. Plumes of sooty smoke from smoldering rice stalks smudge the horizon, the acrid odor permeating the air.

When I ask my students what they consider to be China's greatest problem, the answer is nearly unanimous: too many people. I encounter this problem even in the school, where I struggle to remember the names of three hundred students. One girl icily remarks, "You've asked my name three times today." It's debatable whether a large population is a hindrance or a boon in terms of economic development—prosperous East Asian economies in Japan, South Korea, and Taiwan are backed by some of the highest population densities in the world—but the crowds of China are definitely overwhelming. For relief from the noise and suffocating press of people in the city and around campus, I head out to the hills behind the college, beyond the rice fields, hiking by myself or with students. Mostly, though, the crowds are unavoidable. My students complain to me that Bengbu is a small town; with a million inhabitants, Bengbu ranks a paltry 182nd in urban population among Chinese cities, a mere backwater in this teeming nation.

One of the consequences of China's staggering population density is that meat is largely absent from diets because limited land is available to raise animals and wild game is relatively scarce. Hardly any of my students or teacher colleagues buy meat, partly because it's considered unhealthy, but mostly because it's expensive. When my Chinese students and colleagues taste meat, it's chiefly nibbles or soup bones for flavoring. Most of their calories come from vegetables and, especially, rice or wheat flour in the form of noodles, along with generous helpings of vegetable oils and sugary junk food.

The topic of food weighs heavily on my students' minds. Some of them profess that given a million dollars, they would travel around the world to eat delicious foods or head to Beijing and gorge on Peking duck. To see the way their eyes sparkle like Christmas lights at the mention of Peking duck! In South Korea, people were often categorized on the basis of whether they liked mountains or the sea. In Bengbu, the critical question is: Do you like to eat rice or wheat?

Sidney Mintz, a food anthropologist at Johns Hopkins University, has pointed out that eating unadorned starches is no simple thing. Try downing multiple bowls of white rice, or several boiled potatoes, or a plate of pasta without tomato sauce—it is very difficult for us to eat and digest large quantities of plain starches. Compare that to the ease of sinking your teeth into grilled chicken with a crisp layer of golden skin or a T-bone steak oozing juices. Professor Mintz argues that poor people around the world have historically been relegated to eating flavorless starchy foods, which were made palatable only through the addition of fringe dishes: Think of a thick swirl of spaghetti in a lake of tomato sauce; chilies kicking up corn and beans; or rice with soy sauce, fish sauce, or pickled vegetables.[2] The elites of society, meanwhile, dispensed with the whole business of bland cores and flavorful fringes and helped themselves to meats that were furnished by the laboring masses.

Nor have things changed much; during graduate studies in Los Angeles, I used to pedal my rickety bicycle through Beverly Hills, peering through restaurant windows at elegant people dining on steak, caviar, and sushi, while a pound or so of brown rice—garnished with soy sauce, to be sure—made steady but unspectacular progress through my intestines like a dump truck backing up in a narrow alley.

During this period, I experienced moderate but consistent pain in my lower abdomen area. I had to wake up to use the bathroom at the same time every night. I went to the university health clinic to get an assessment. I didn't have high hopes, since the doctors and nurses there had been previously baffled by my symptoms. However, the pain was affecting my ability to concentrate, and I was worried that it augured something serious.

The nurse whom I saw wasn't too worried. "Look," he said soothingly, like a parent plastering a Band-Aid over a child's scraped knee, "the urge to urinate at night is perfectly normal. It's very common. I have to get up every night myself."

But I didn't have any of these symptoms a year ago. Maybe the problem was related to my bike seat. No, the pain continued when I stopped riding. At the time, I was famous among my colleagues for bringing lunch boxes that were packed with brown rice. A few months after the visit to the clinic, I dropped by my parents' home in Canada and learned that they, too, experienced discomfort from eating brown rice; in fact, they had stopped eating it. Within days of going off brown rice, my pain disappeared. I crowed triumphantly to my mother, "That was it! I don't have to go to the bathroom every night anymore. It was the brown rice!"

The pain was likely aggravated by hernias incurred while struggling with heavy (for me) weights in the gym, trying to add some Schwarzenegger-like bulk to my toothpick frame—when you hit the beach in L.A., you couldn't afford to look like some anemic geek who spent all his time in a cubicle turning textbook pages. Yes, I should have soaked the brown rice for a few hours before cooking, which would have made the rice softer and easier to digest, but as a grad student with a dissertation to complete, it was taxing to remember my name, let alone remember to soak the brown rice.

Considering the blandness of starchy foods and the difficulty of digesting them leads to one of the most important questions concerning the history of humanity: Why did humans give up hunting and gathering for sedentary agricultural life? After all, hunting and gathering appears to be much more rewarding than the backbreaking labor of a farmer. On top of that, the hunter-gatherer gets a pretty good meal out of wild game and assorted veggies, fruits, and nuts, while the farmer gets . . . well, a lot of starch that needs to be doused with salt, sugar, oil, or spice to make it palatable.

The transition from mobile hunting and gathering to sedentary agriculture took place in thirteen to twenty-four different locations around the planet, starting around twelve thousand years ago, and proceeded in fits and starts for several thousand years. Many explanations have been proposed, but none have gained widespread acceptance among archaeologists or other researchers. One theory holds that hunter-gatherer populations expanded and created strain on local food supplies until it was necessary to give up the leisurely life of the hunter-gatherer for the nutritional deficiency and toil of farming life. Another major set of theories focuses on climate change. Around twelve thousand years ago, the climate ceased to fluctuate and became cooler and drier, and more atmospheric carbon dioxide was available, which could have made growing crops feasible for the first time.[3] However, it turns out that some societies increased in population only after they took up agriculture, or switched to farming while their populations were declining. Moreover, early agriculture seems to have occurred in areas that had plentiful food, rather than food scarcity.[4]

Another possibility relates to the fact that humans cannot consume more than 35 percent to 40 percent of calories in the form of protein, due to the accumulation of toxic levels of ammonia and urea as byproducts of digesting and metabolizing protein. Although protein must be consumed, fat and/or carbohydrates must provide the bulk of calories. The end of the last ice age twelve thousand years ago caused forests to take over grasslands and thus created habitat stress on large mammal populations, but humans hungry for fatty meat certainly also helped the megafauna to their demise. Moreover, the appetite of the hunters for fatty meat guaranteed that no new large mammals evolved to take the place of the extinct megafauna. After humans migrated from Africa and entered Australia, New Zealand, Tasmania, North and South America, Madagascar, Japan, and other landmasses, the lumbering feasts of meat—giant marsupials, giant deer, giant flightless elephant birds, giant lemurs, giant beavers, and more game—were the first to go extinct, followed by their smaller, more nimble, and less fatty relations. Large mammals were sometimes able to survive in dense forests (like those in the Amazon and Southeast Asia) or the coldest, most forbidding regions of the world, such as the New World Arctic. Big game also persevered in Africa; due to the long history of coevolution with bipedal (walking upright on two feet) hunters on the continent, animals may have evolved to be wary enough to survive Neolithic weaponry.[5]

As climate change and hunting pushed the great mammals, lizards, and flightless birds to extinction, hunter-gatherer groups could compensate for the loss of animal fat by subsisting on leaner game animals. For example, in Southwest Asia, large game like wild cattle, deer, and wild boar steadily disappeared from diets starting around thirteen thousand years ago, replaced with smaller animals like mountain gazelles, tortoises, hare, and partridge. Not only did people make use of progressively smaller and leaner animal species, but they also resorted to catching younger gazelle and intensively harvesting gazelle marrow, as well as collecting grass seeds.[6] Some groups in Southwest Asia became more sedentary as they hunted out their big game, but others may have become more mobile in an increasingly desperate attempt to get more.[7]

Thus the critical factor behind the transition to agriculture may have been the loss of big fatty prey. (Insects can also be fatty, but they take more effort to gather per calorie attained, and their chitin exoskeletons may also present problems, as discussed earlier.) Frustration with increasingly lean diets could have led to the adoption of sedentary agriculture and domestication of animals as a last resort. People's health was compromised by the new diet: Diminution in height and the appearance of dental cavities have been noted in the remains of post-agricultural-revolution peoples.[8] But that wasn't the worst case. If suitable plants and animals were unavailable, the increase in human population was checked, forcing tribes to live or starve according to the availability of fluctuating food supplies, until migrants from other regions introduced novel domesticated animals or plants.

Plant foods are today heralded as healthy fare, but people in traditional societies generally did not favor them, and for good reason. Consider the fate of the infamous Burke and Wills expedition, a scientific caravan that departed Melbourne in 1860 with the intent of exploring and crossing the Australian interior. The retinue boasted food sufficient for two years and sixty gallons of rum (to revive the camels)—all told, about twenty tons of supplies. However, after several months of mishaps and errors of judgment, three men—Robert O'Hara Burke, an Irish soldier and police officer; his second-in-command, William John Wills, a young English surveyor; and John King, an Irish soldier—found themselves stranded at Cooper Creek, hundreds of miles from Melbourne, with no pack animals (some of the camels had been eaten) and dwindling food supplies. Suffering from malnourishment and exhausted, the three men traded their sugar with native Aborigines for fish, beans, and the spore-like fruit of the nardoo fern. The Aborigines ground nardoo to make a paste and bread that were valuable during drought conditions, but the explorers may have neglected to roast, sluice, or winnow the spores as the Aborigines did. Doing so would have purged the nardoo of thiaminase, an enzyme that destroys vitamin $B_1$. A person lacking vitamin $B_1$ is debilitated by beriberi, a condition characterized by paralysis, weight loss, and loss of feeling in the extremities. Even though the men were able to consume four pounds of nardoo a day, they steadily lost strength. After weeks of wasting away, Burke and Wills died at Cooper Creek; a rescue party eventually recovered a seriously weakened King.[9]

The Burke and Wills debacle is usually portrayed as an example of cultural incompetence, since the explorers relied heavily on goods and technological might, while the Yandruwandha Aborigines were able to survive in the same area using the accumulated wisdom of their forebears. However, even the Aborigines used nardoo only as an emergency food. Consider the situation from the nardoo plants' point of view. Like a squatter squaring off with a bulldozer, if you put down roots in a patch of soil, intending to live there for the rest of your life, you'd put up a rocking good fight the moment anyone tried to browse on your limbs, prune your flowers, pull out your roots, or nibble on your immature seeds. Making the best of their immobility, plants discourage predation with an impressive battery of defensive compounds. A raised middle finger or a portrait of Che Guevara may be the conventional symbols of defiance to many, but a plant would be just as true to the spirit of resistance.

Apart from their fruits, plant parts are designed to be unpalatable through physical barriers or chemical warfare. We can group plants into six categories based on their effects when consumed:

- *Enemies*: plants that should never be eaten. These include *assassins*, plants whose toxins we deliberately employ as means of carrying out murder, torture, or punishment.
- *Doppelgangers:* plants that poison us because we mistake them for palatable plants.
- *Sorcerers:* plants that we regularly use as medicines but that harm us when we accidentally overdose on them.
- *Werewolves:* plants that are safe to eat at certain stages in their life cycles and dangerous at other life stages.
- *Fallbacks:* plants that may be eaten as a temporary resort but are not suitable for long-term consumption.
- *Comrades*: plants that are suitable for long-term consumption when properly prepared.

Much of the confusion today over which plants to eat results from people indiscriminately lumping plants into the comrades category. However, just as in human relationships, not every plant that we meet is suitable as a long-term mate. Our ancestors were far more likely to place plants into the sorcerer or fallback category and instead regard animal foods as their true companions. People in traditional societies preferred to cook every vegetable, rather than eating them raw: Cooking is the best means of neutralizing plant defensive toxins in a diet that is largely based on vegetables, as well as unlocking edible calories in curmudgeonly plant tissues. White starch—from white rice, wheat flour, boiled potatoes, and so on—is highly valued (after meat, in any case) in traditional societies because in the long run, it is least likely to harm human eaters. Despite the gush of enthusiasm that nutritionists often profess for plant foods, citing antioxidants, vitamins A and E, fiber, polyunsaturated and monounsaturated fatty acids, potassium, or the absence of cholesterol and sodium in plants, the truth is that none of these properties have so far demonstrated conclusive benefit in well-nourished populations. What makes a particular food or cuisine healthy is whether or not it supplies the nutrients that our bodies evolved to require. Plant foods became an increasingly major force in our diets as our big animal prey dwindled around the world because people learned how to process, cook, and selectively breed the new plant foods.

Let us consider now a few examples of each of the plant categories, to reach a better understanding of the mysteries behind plant foods, and to illustrate that even common plants may have surprising properties.

**ENEMIES AND ASSASSINS:** Some plants are plain poisonous but were employed anyway, as tools for punishment, murder, or suicide. A cruel punishment for Jamaican slaves was rubbing their mouths with the cut stalk of dieffenbachia (now a common houseplant), which caused painful swelling of the oral mucous membranes and rendered them unable to speak; hence its nickname of dumbcane. The extract was used as an ingredient in preparing arrow poisons by natives in the Amazon.[10] Another example is the castor oil plant, planted as an ornamental. Castor oil is widely employed as a highly effective laxative, but the seeds contain ricin, one of the most potent poisons known. In 1978, the broadcaster Georgi Markov, a Bulgarian exile in London, was waiting for his bus to take him over the River Thames to his job at the BBC when he was bumped in the leg with an umbrella. The symptoms of his agonizing death, three days later, were consistent with assassination by ricin poisoning. The poison capsule that had tipped the umbrella and that had been injected under his skin was no larger than the head of a pin.[11]

**DOPPELGANGERS:** Sometimes we eat noxious plants because of mistaken identity. The toxin from water hemlock, a plant sometimes confused with wild parsnip, wild carrot, wild celery, artichokes, sweet potatoes, or sweet anise, triggers spasms violent enough to make people bite through their tongues and smash apart their teeth.[12] Another example of a doppelganger is meadow saffron, which may pass for onion; eating meadow saffron triggers thirst, diarrhea, stomach pain, delirium, and death in half of all cases. It may take up to three days for death to deliver merciful deliverance.[13]

**SORCERERS:** Certain plants may be fatal to consume, but we attempt to employ them in small doses as medicines. Dieffenbachia, the houseplant mentioned above, was chewed by males in the Caribbean Islands to achieve temporary sterility lasting up to two days. Bitter melon, a fruit commonly found in Asia and in Asian grocery stores elsewhere, is anecdotally reported to improve symptoms of diabetes. In 2010, an Indian scientist reportedly died from drinking a particularly bitter concoction of bottle gourd and bitter gourd juice, a regimen that he had maintained for four years. His wife also drank the juice, but survived after vomiting blood and experiencing severe diarrhea.[14] Bitter melon, cucumbers, and squashes concoct a bitter compound called cucurbitacin to protect themselves from insect and fungal attack. Although domesticated versions of these plants were bred to reduce their bitterness, when cucurbitacin is present in high concentrations, the bitter taste—gardeners are familiar with the bitterness of cucurbitacin in homegrown cucumbers near the stem end—normally compels people to stop eating before they become ill.

**FALLBACKS:** Sometimes, poor people are forced to subsist on noxious plants out of hunger and poverty. Lathyrism, a disease that leads to back pain and paralysis of the lower limbs, results from prolonged diets of grass pea.[15] Under conditions of poverty, famine, and interruption of agricultural work, people consume the hardy grass pea plant as a last-resort food item. Lathyrism debilitated thousands of people in northern India, and outbreaks occurred during hardships such as the Spanish Civil War (1936–39), among Romanian Jews confined to concentration camps, Greeks besieged by the Germans during World War II, and German inmates in France just after the close of World War II.[16] Lathyrism may not just be a problem in Europe and Asia: Christopher McCandless, the young American itinerant whose life was recounted in the popular book and movie *Into the Wild,* may have died from lathyrism incurred by eating wild-potato seeds while attempting to live off the land in the Alaskan woods.[17]

**WEREWOLVES:** Some plants are only troublesome at certain stages in their life cycles. In the fall of 1978, up to three hundred boys at a southeast London day school sat down for lunch, choosing from a menu of potatoes, steak pie, gravy, cabbage, tinned carrots, and dessert of apricot and syrup sponge pudding, with or without custard. By eight o'clock that evening, seventy-eight of the boys began to vomit and experience severe diarrhea and stomach pain. Seventeen boys were taken to the hospital; they developed fever, and their feces turned green. Three boys fell unconscious, and two spoke gibberish when they regained consciousness. Fortunately, by the eleventh day, all the boys had recovered enough to be discharged. The one food that all the stricken boys had eaten in common: potatoes. Domesticated potatoes have been bred to reduce the steroid alkaloid solanin to palatable levels, but tubers that are exposed to sunlight (thus making them vulnerable to being eaten) and turn green, or that have been attacked by disease or left to spoil, may produce dangerously high levels of solanin. Since solanin tastes exceptionally bitter, fatal poisoning by potatoes is uncommon, but it does occur, such as during the Korean War, when large segments of the North Korean population were reduced to subsisting on spoiled potatoes. Investigation into the London case suggested that the affected boys had consumed potatoes from an old bag that had been left over from the term prior to the summer.[18]

**COMRADES:** Finally, we arrive at our plant BFFs (best friends forever), the vegetables, legumes, and cereals that we find in grocery stores, in our garden plots, and on our farms. These are the foods that most nutritionists recommend that we heap onto our dinner plates, lightly cooked or processed, if at all. Traditional societies certainly valued plant foods like these, but they were careful to process and cook them in ways that reduced their harmfulness. The chief benefit of our companion plant foods is that they do not poison us—not outright, anyway. Let us consider some of the defensive compounds that our everyday plant foods attempt to deploy on predators like ourselves.

Some defensive compounds cannot be reduced by cooking. For example, celery, parsley, and parsnip stock up on furanocoumarin, a compound that protects plants against insects but also can cause skin rashes in people who handle these plants (though eating celery does not present this problem). The skin rashes can be a hazard for field workers and are worsened when exposed to sunlight. Ironically, breeding for celery varieties that are more resistant to insects or fungi may inadvertently ramp up the concentration of furanocoumarins.[19] Saponins, a defensive toxin deployed in chickpeas, soybeans, beans, peanuts, spinach, and asparagus (and also in sea cucumbers), are soaplike compounds that create a bitter taste and irritate mucous membranes. Saponins are toxic to cold-blooded animals such as insects and fish and are used around the world as fish poisons. Normally saponins cannot pass through the gut wall of humans. However, if for some reason saponins enter the bloodstream, such as through an injury to the gut, they can cause rupturing of the blood cells (hemolysis). Symptoms of saponin poisoning include dizziness, headaches, chills, irregular heartbeat, convulsions, and coma. Like furanocoumarins, saponins are fairly resistant to cooking or most other techniques of food processing, with the exception of fermented foods like tempeh (Indonesian fermented soybean), which have substantially reduced saponin content.[20] Isoflavones, compounds that resemble and mimic the properties of estrogen, are produced by soybeans and to a lesser extent by other plants from the legume family, including alfalfa and clover. Animals that browse too much on isoflavone-rich plants, such as ewes feeding on clover, can become sterile, due to the disruptive hormonal effects of isoflavones. Soy-based formulas can interfere with steroid metabolism in infants. Like furanocoumarins and saponins, isoflavones are resistant to cooking.

In other cases, cooks in traditional societies learned how to make a good meal out of well-defended plants through ingenious food preparation techniques. Legumes (beans, soybeans, lentils, chickpeas, etc.) fortify their seeds with lectin compounds that cause gastrointestinal distress when consumed in large quantities, as well as growth reduction and liver damage.[21] Protease inhibitors are a class of defensive compounds manufactured by legumes, cereal grains, and potatoes to prevent plant predators from being able to digest their foods. Hard-pressed peasants figured out that concentrations of lectins and protease inhibitors are reduced through food preparation methods such as sun-drying, pan-frying, deep-frying, roasting, soaking, boiling, and fermentation.[22] Cassava is problematic due to cyanide compounds; populations in tropical countries that rely on cassava as a staple may suffer from cyanide poisoning, goiter, or neurodegeneration. Traditional methods of making cassava safer to eat include sun-drying, soaking, grating, and roasting. Lima beans, sorghum, and bamboo may also induce cyanide intoxication. Clever tricks to reduce cyanide content include grating, chopping into smaller pieces, drying, boiling, prolonged submersion in warm or hot water, steaming, roasting, and fermentation.[23] (Boiling should be done uncovered, to allow cyanide gases to escape completely.)[24]

There are also plant parts that are not built specifically for defensive purposes but can still damage the health of predators. Like an attractive best friend who mesmerizes your potential dates, phytates are storage forms of phosphorous that bind to minerals and therefore have the tendency to deplete our bodies of essential minerals such as calcium, magnesium, zinc, and iron. Soybeans, beans, cashews, sesame seeds, pistachios, chickpeas, peas, apples, eggplants, tomatoes, and papayas contain phytates. Oxalates, which may also steal your calcium and other minerals and are a risk factor in developing kidney stones, are abundant in spinach, okra, chocolate, couscous, whole-meal rye, whole-wheat bread, durum wheat, and especially wheat bran. Dehulling, soaking, cooking, and sprouting are common food preparation methods that reduce phytate content, while oxalate levels in foods are reduced by dehulling, boiling, steaming, baking, breadmaking, and fermentation.[25] In other words, our relationships with our plant food comrades, even those that have been with us for a long time, require a lot of work to maintain on good terms.

The great irony about plant foods is that the more we reduce their harmful by-products and chemical defenses, the more sugarlike they become, and thus the more we increase our risk of acquiring chronic diseases like type 2 diabetes and gout. The ultimate reason for this double-edged character of plant foods is that they are not our original food source; we don't have the specialized digestive systems or teeth that dedicated herbivores like gorillas and cows possess to grind and digest large quantities of unprocessed plant foods, and must make do with a series of ingenious culinary workarounds to render plant foods suitable eating.

But such ingenuity! The staples that we grow up with become enshrined in our hearts, and the plant that has been arguably closest to Westerners' hearts for several millennia has been wheat. In the Lord's Prayer, Christians recite: "Give us this day our daily bread." Wild wheat was gathered at least as far back as 17,000 BC. The great virtue of wheat is that it contains starch, an easily digestible carbohydrate, and gluten protein, which is sticky and can be leavened with yeast to make bread (rice lacks gluten and therefore makes poor bread); as mentioned, the process of fermenting wheat and turning it into bread reduces harmful oxalate levels.[26]

However, wheat—and gluten in particular—has become the scorn of a rising movement, blamed for instigating a plethora of diseases. Celiac disease is an autoimmune intestinal disorder triggered by the form of gluten protein found in wheat, barley, rye, and closely related cereals; it currently affects 1 to 2 percent of people in Western countries. The symptoms of celiac disease usually take months or years to develop after exposure to gluten. Children with celiac disease may eventually exhibit anorexia, lack of energy, pale skin, growth retardation, delayed puberty, or rickets, while adults with celiac disease may experience symptoms including diarrhea, nausea, vomiting, stomach pain, flatulence, and weight loss.[27] Celiac disease is also prevalent in North Africa, India, and the Middle East. Wheat, barley, and rye have been major food sources for thousands of years in the areas where celiac disease is most common, so why has natural selection not curtailed the frequency of the genes underlying celiac disease?[28]

One possibility is that the cereals that trigger celiac disease have not been consumed by us for a long enough time for evolution to do its work. According to this argument, the three hundred or so human generations that have been exposed to these cereals were not affected enough by celiac disease for it to have diminished the ability of these people to bear children. The problem with this argument is that celiac disease is a serious disorder and would have harmed a person's reproductive prospects in the days before medical treatment and gluten-free diets were widely available.

An alternative argument is that the genes that promote celiac disease may somehow give people better health in other ways. When scientists scanned gene databases for patterns in celiac disease, they found that some of the genes that lead to celiac disease increased in frequency between 1,200 and 1,700 years ago, just when human dependence on cereals should have pushed the genes to obscurity. The key is that these same genes are also involved in protecting people against bacterial infections. In other words, celiac disease may be a double-edged disorder, which confers bacterial protection but makes bearers of these genes vulnerable to gluten poisoning.[29]

However, celiac disease is increasing rapidly, and not everyone who has a genetic predisposition toward celiac disease develops it.[30] Changing genes cannot be a complete explanation. Something crucial in the environment must have altered as well. Scientists have recently observed that birth via Cesarean section may increase the risk of celiac disease, perhaps due to the lack of transmission of the mother's intestinal bacteria to the baby; overuse of antibiotics may similarly reduce intestinal bacteria and increase the risk of celiac disease.[31]

Other disorders besides celiac disease have been blamed on the consumption of wheat. For example, wheat allergies have become more prominent (the topic of food allergies will be discussed in a later chapter). Other gluten reactions seem to involve neither autoimmune (as in celiac disease) nor allergic mechanisms and are currently lumped under the label "nonceliac gluten sensitivity," or more often "gluten sensitivity." Commonly reported symptoms of gluten sensitivity include headaches, "foggy" states of mind, fatigue, depression, bone or joint pain, muscle cramps, leg numbness, and weight loss. People with gluten sensitivity believe that the symptoms improve when gluten is removed from the diet, but many medical practitioners and doctors are skeptical because clinical experiments have not demonstrated any symptoms from gluten consumption so far. The link with gluten may be a nocebo (i.e., negative placebo) effect, purely in the mind. The perceived problems with eating wheat products may stem from other chemical components that are present alongside gluten; for example, there has been a surge of interest in the study of short-chain sugars (also known as FODMAPs, for fermentable oligo-, di-, and monosaccharides and polyols) that ferment quickly in the intestines, causing bloating, gas, gastroesophageal reflux, and diarrhea, and may be the true cause of gluten sensitivity. FODMAPs are extremely widespread in contemporary Western diets, in the following chemical forms and foods:

- *free fructose* in apples, cherries, mangos, pears, watermelons, asparagus, artichokes, sugar snap peas, honey, and high-fructose corn syrup
- *lactose* in milk, yogurt, ice cream, custard, and soft cheeses
- *fructans* (fructose chains) in peaches, persimmons, watermelons, artichokes, beetroot, Brussels sprouts, garlic, leeks, onions, peas, wheat, rye, barley, pistachios, legumes (beans), lentils, and chickpeas
- *galacto-oligosaccharides* (short chains of galactose sugars) in legumes, chickpeas, and lentils
- *polyols* (sugar alcohols) in apples, apricots, pears, avocados,

blackberries, cherries, nectarines, plums, prunes, cauliflower, mushrooms, and snow peas[32]

Intestinal discomfort may therefore result not simply from eating
too much wheat (or gluten) but rather from eating too many sugary
foods, including factory-made sweetened breads as well as fruits and
fructose-containing foods. Like celiac disease, FODMAPs may lead
to intestinal discomfort when intestinal bacteria populations are altered through the overuse of antibiotics. Another consideration is that
moderate exercise can help to reduce gastrointestinal disorders such
as irritable bowel syndrome and constipation, whereas too much vigorous exercise can exacerbate gastrointestinal disorders like reflux,
heartburn, diarrhea, and gastrointestinal bleeding.[33] Overall, plant
foods are best prepared and eaten in traditional ways—grated, steamed,
roasted, fermented, and so on—rather than served raw, and comple-
mented by a lot of moderate exercise like walking and the avoidance of
particularly sugary foods, which will help alleviate intestinal discomfort from FODMAPS-induced gas.

When I first arrive in China, I eagerly try the street food in the student
ghettos surrounding Bengbu College, but the fare—wheat wraps, ver-
micelli soups, barbecue skewers—is disappointingly oily and spicy. I
know there must be better food out there, but the students generally
can't afford to eat off campus, and I don't speak enough Mandarin to
befriend the teachers who can't speak English. As a result, I spend most
of my evenings at the track, trying to jog while weaving around chatting students and families, or at the gym, playing basketball, badmin-
ton, table tennis, and volleyball with students and teachers. After
Christmas, the English Department desperately needs volleyball play-
ers to help trounce the Nutrition Department, the latest incarnation
of the annual teachers' competition. Although I am technically a member of the International Relations Department, I had played setter
and captain on my high school's junior volleyball team, so through
bureaucratic sleight of hand, I become a bona fide member of the

English Department. It's going to be a big match, I'm told over and over by jittery colleagues. The day of the big game, I get psyched and participate in a super-vigorous warm-up session. Unfortunately, I have been experimenting with low-glycemic diets of barley, oats, millet, beans, and other indigestible food and warm up so hard and have been out of volleyball for such a long time that I am famished and exhausted by the time the game starts. As the students crowd the sidelines and yell out coordinated cheers, I fan at the ball a few times and botch some easy sets. We squander the big match. My colleagues are crushed. However, I am told there will be another set of games the next day, a chance for possible redemption.

Before heading to the match the following afternoon, I pull out a frozen banana from the freezer and microwave it to a steaming calorie-rich mush and pop an obsidian-black "thousand-year" cured and salted duck egg in my mouth. I bound to the gym like a 150-pound Vietnamese version of the Incredible Hulk. The English Department students have mostly given up on us and gone home for the long-weekend holiday, but I pound the ball and scream as if it's an Olympic showdown. After losing our first match, we win our second match against the Biology Department, salvaging pride and second place overall. The English Department has secured its reputation for another year.

I am invited to a celebratory banquet at a fancy restaurant in Bengbu. Cigarettes are passed around, hard liquor poured out. This evening is about slapping backs, shaking hands, pouring drinks for senior colleagues, making toasts—all the schmoozing needed to smooth out office politics for the upcoming months. I am toasted for my efforts on the volleyball court, but my attention is riveted by the platters of food being slowly spun around the table: honeyed slices of lotus, crispy fried carp, tender marinated beef quivering with fat, savory pork ribs, a mob of other delicacies, every dish a winner. My salary at the college is $800 a month, of which $400 is siphoned off to service my student loans, which makes me poor even by Chinese standards. I'll never see the likes of this food again. I'm feted with rice wine until my head can barely stay level, but like a man just rescued from weeks of being stranded in the desert, I continue to obsessively pick at the remnants of the dishes while the teachers fervently schmooze.

Sumptuous meaty feasts like this are a rarity in China. When two of my students invite me to visit their homes, two-story concrete dwellings set in sparsely forested countryside, some gamey chicken and bits of pork are served, but the mainstays on the table are tofu, eggs, peas, tomatoes, peanuts, and greens, fried in lard or, increasingly, vegetable oils, and, of course, white rice, with green tea or rice wine. Due to the lack of red meat and dairy, this type of cuisine is likely to delay the onset of chronic diseases like breast and prostate cancer, but the oily sautéed greens and white rice are likely risk factors for obesity and diabetes, respectively. When my students ask me what I think of Chinese cuisine, I tell them I like traditional Chinese cuisine, when lard was used to lightly sauté vegetables, rice was hand-milled and accompanied by yams or grains like barley, and insects, fish, and frogs supplemented the meat. Unfortunately, such food is quickly becoming a distant memory. Now a bullet train thunders through the countryside around Bengbu, cars, tractors, and motorcycles have replaced walking, old-fashioned lard has been replaced with cheap vegetable oils like soybean and corn oil, rising incomes are used to purchase meat and milk, and the rice is white as snow. In a country obsessed with modernizing, it will take another generation before Chinese people realize how damaging to health these dietary and lifestyle changes have been. Then the Chinese will start to look back to the ways that their ancestors once lived and ate.

# ELIXIRS

Many children are not consuming recommended servings of dairy foods. . . . Dairy foods provide essential nutrients needed for body maintenance and protection against major chronic diseases.

—GREGORY D. MILLER, JUDITH K. JARVIS, AND LOIS D. MCBEAN, *Handbook of Dairy Foods and Nutrition*

Cow's milk consumption is a major health hazard and should be recognized as a promoter of most common chronic diseases of industrialized countries.

—BODO C. MELNIK, *"Milk—The Promoter of Chronic Western Diseases"*

Three liquids have had an enormous impact on human health over thousands of years: water, alcohol, and milk. Tea and coffee have also been important beverages in parts of the world, particularly in terms of their economic impact in recent centuries, but because scientists are still largely unclear about the long-term health effects of drinking tea or coffee, we won't delve deeply into these beverages. I've had long discussions with friends about the mood swings associated with coffee/caffeine dependence, but given the extreme irritability such discussions engender (especially as the time since the last caffeine hit drags on), I've learned not to frown or roll my eyes as my friends fork over their cash to the coffeehouse giants. In any case, according to one recent large American survey (more than 250,000 men and more than 170,000 women between the ages of fifty and seventy-one), drinking six cups of coffee or more a day lowers the risk of dying by 10 percent (men) or 15 percent (women); so perhaps my coffee-addicted friends have justification for frowning and rolling their eyes at *me*. The most obvious explanation for why drinking coffee cuts the risk of death is that the alertness accompanying caffeine consumption decreases the chance of getting into a fatal accident, but coffee drinkers also seem to have better chances of avoiding heart disease, respiratory disease, stroke, diabetes, and infections, which suggests that some other compound in coffee besides caffeine (like polyphenol antioxidants) might confer health benefits. Similar reductions in mortality have been obtained in Japanese studies on coffee drinking. Scientists haven't noted any similarly substantial health benefits from drinking tea so far.[1]

Compared to coffee or tea, to say nothing of milk or wine, water seems plain and uncool. The old adage of drinking eight glasses of water a day seems almost quaint now, but there is a lively debate around the merits of drinking water, or at least around drinking certain types of water. About fifty years ago, doctors and scientists first noticed that in regions where people drank water that was "harder" (i.e., water having less acidity, leading to more bathroom scum and less soapy water), people tended to suffer from less heart disease and live longer. The first candidate for a life-boosting, blood-pressure-lowering chemical element was calcium. However, as researchers accumulated more data, they began to realize that calcium was probably not the responsible element, and attention shifted toward magnesium. Inadequate levels of magnesium could make heart rhythms more irregular, worsen the lipid profile and insulin control, and lead to more plaque buildup in arteries, all of which could exacerbate the risk of heart disease. Normally, if one thinks of magnesium at all, it's supposed to come from a diet rich in vegetables, fruits, and nuts, but magnesium ions in naturally hard water are likely more easily absorbed by the body than the magnesium found in food or vitamin supplements. The importance of magnesium in water is especially disconcerting for parts of the world where there is increasing reliance on water sources other than groundwater, such as recycled wastewater and desalinized seawater. When I have the money, I like to drink bottled waters from Italy that have a lot of dissolved minerals. Such water has a superb taste and a lot of magnesium, perhaps the way nature intended it, but I understand that from an environmental perspective, not to mention my poor wallet, it makes little sense to import drinking water from other continents.[2]

Drinking water straight from a lake, stream, or puddle—great for acquiring useful minerals but also bothersome parasites—was certainly the chief drink of our ancestors. However, other beverages were even more valuable, for their powers to make us bold and imbecilic, or strong and tall, like magic potions out of a fairy tale.

Our distant ancestors from millions of years ago would have been familiar with the taste of alcohol in fermenting fruits, but the ability to manufacture any sizable quantities of alcohol had to await the development of agriculture. The earliest evidence so far of the use of alcohol has been found in China, dating to around seven thousand years ago. Alcohol doesn't preserve well because it evaporates quickly (which is why leftover wine is discarded), so any evidence about the use of alcohol in ancient times is necessarily indirect, but clever archaeologists have put together the following scenario. The Chinese domesticated rice at least nine thousand years ago, and rice is a good start for making alcoholic brew. However, making alcohol requires yeast, and yeast doesn't grow on rice. Nowadays, a common method of making alcoholic beverages in Asia is to grow mold on blocks of rice or other cereals; the mold is introduced by chance through insects or by parachuting down from old ceiling rafters. The ingenious solution the ancient Chinese seem to have hit upon to obtain their alcohol was to mix hawthorn fruit and honey (both of which harbor yeast) with rice (to provide the yeast with fuel for generating ethanol), a combination that would have begun fermenting in a few days in a warm climate.[3] Researchers have found evidence of beer-making using fermented barley in the Zagros Mountains of Iran dating to more than five thousand years ago. Around the Mediterranean, the idea of using grapes to make wine, likely mixed with medicinal herbs, was present from at least 3,000 BC in Egypt.

While there is considerable evidence that humans started to make alcohol in impressive quantities after the advent of agriculture, why we enjoy drinking it is a surprisingly controversial topic. Some scientists believe that animals, including humans, have a deep instinct to seek out ethanol, as a marker of valuable energy-dense fruits and nectar. Indeed, alcoholic beverages were an extremely important source of energy in some preindustrial diets; the daily ale consumption of an Englishman in the sixteenth century could exceed one gallon. Ethanol is a calorie-dense product: One gram of it packs seven calories, instead of the usual four that come from sugar and almost as much as the eight calories that fat delivers.

However, when frugivores like birds and fruit bats are allowed to choose between ripe fruit and alcohol-soaked rotten fruit, or between foods with varying quantities of alcohol, they almost always settle for the nonalcoholic choice.[4] My uncle introduced me to beer when I was a kid, and it tasted much as my older brother had described it: horse piss. Considered from the perspective of ethanol-spewing yeast, this makes sense. Yeast and other frugivores, like bacteria, birds, and humans, have conflicting interests: We all want the fruit. When I was in grade school in Ottawa, a common tactic wily children employed to defend their desirable snacks was to spit on them, which swiftly rendered the contested item personal property. Likewise, yeasts like *Saccharomyces* have a knack of transforming fruit sugars into ethanol, because ethanol is toxic to frugivorous bacteria and vertebrates. That suits the yeast fine, because it can then employ its alcohol enzymes to convert the ethanol into usable sugar for its own leisurely consumption.[5]

When cedar waxwing birds gorge on alcoholic berries and then slam into windows and telephone poles—"If You Drink, Don't Fly," quipped a scientific paper title—it is more likely they are famished at the end of winter and less likely they are looking for a last drink for the road.[6] (Some shrews that feed on alcoholic nectar ingest great quantities of alcohol for their body weight but show no signs of intoxication, which may mean that they have evolved physiological mechanisms to process the high alcohol content of their customary diet.)

Paradoxically, among humans in modern times, study after study has found that moderate alcohol consumption helps fight coronary heart disease. The customary finding is that around two drinks (such as a glass of wine, a can of beer, or a shot of spirits) a day for men or one drink a day for women cuts the risk of coronary heart disease, ischemic stroke (when the blood supply to the brain is cut off), and sudden death. No one has discovered exactly why alcohol has this effect, but alcohol consumption results in higher levels of healthy HDL cholesterol and of a protein that shuttles cholesterol out of arteries (ApoA-I), and in lower levels of a protein (fibrinogen) that may exacerbate the risk of cardiovascular disease, possibly through increase in clotting. Many people swear by the rosy blessings of red wine, but the advantages of alcohol may apply to alcohol in general.[7]

That being said, the benefits of booze accrue principally to people at risk of coronary heart disease, which is to say people in industrialized countries over the age of forty. Among young people, the causes of death tend to be the kind of drama that alcohol consumption aggravates rather than ameliorates: accidents, suicides, and homicides. (In developing countries, infectious diseases are the main killers, and thus tipping back a few drinks a day has little or no protective effect on mortality.) Indeed, heavy alcohol consumption (especially binge-drinking) has well-documented health risks, such as an increased risk of cirrhosis of the liver, hemorrhagic stroke (leaking or burst blood vessel), cancers of the upper digestive tract, and metabolic syndrome, a cluster of health conditions including high blood pressure, diabetes, and obesity.

Ethanol is a carbohydrate, like glucose, and delivers energy to the body. However, there are crucial differences in the way that ethanol and glucose are metabolized. Glucose has a deep evolutionary history in humans and other forms of life and is therefore recognized and welcomed throughout the body's tissues. By contrast, ethanol does not evoke an insulin response and stealthily penetrates the liver, potentially resulting in liver damage, insulin resistance, and metabolic syndrome, as noted above.[8]

Moreover, it is worth bearing in mind that the harmful effects of alcohol are much more pronounced in women than in men. For a given amount of alcohol consumed, blood concentrations of alcohol in women tend to be higher because of the smaller average body size of women compared to men. A higher proportion of body fat in women also reduces the water space in women's bodies, again increasing the blood concentration of alcohol in women relative to men. Among people less than fifty years of age, women tend to have less gastric alcohol dehydrogenase (enzyme) activity than men; with less alcohol broken down in the stomach, a large amount of alcohol is able to enter the bloodstream. Liver injury is more frequent and progresses faster among women than among men with similar histories of alcohol abuse. When exposed to alcohol, women have higher toxic acetaldehyde levels than men. This is why around one drink per day is considered to be moderate alcohol consumption for women, versus two for men.

Since alcohol is so detrimental to human health, one might expect that humans have adapted genetically to handle exposure to alcohol. Indeed they have—but in surprising fashion. The alcohol dehydrogenase (ADH) gene helps us convert alcohol to acetaldehyde. Acetaldehyde is usually soon broken down by another enzyme (acetaldehyde dehydrogenase) and glutathione to harmless acetate, but if the body is overloaded with acetaldehyde, the liver cannot make enough glutathione to match the demand, and thus toxic acetaldehyde builds up. Around ten thousand to seven thousand years ago, a variant of the ADH gene began to appear in human populations and became especially prevalent in East Asia. Oddly, this gene variant leads to greater production of toxic acetaldehyde and consequently the flushed faces, headaches, and hangovers that commonly occur when East Asians drink booze. This type of reaction discourages the ADH gene bearers from overdrinking and hence protects them.

Incidentally, the drug disulfiram, used to treat alcoholism, results in headaches and vomiting after ingestion of alcohol, thereby strongly discouraging further drinking. In my own family, my dad and my younger brother both get flushed faces from drinking alcohol and avidly dislike wine and beer; my older brother and I do not get these symptoms and enjoy our liquor very much. Indeed, scientists have observed that people with the protective ADH gene variant are far less likely to become addicted to alcohol. Fittingly, this gene variant increased in frequency where rice cultivation first arose and where rice wine—and bouts of exaggerated human drama—came soon after. By contrast, in other parts of the world, the farther one travels from East Asia, the rarer this protective gene becomes (it is very rare among peoples of the United Kingdom and the New World), which suggests that alcohol drinking has had a shorter history in these places, as well as a potentially more devastating effect on health, given the lack of protective genes. That being said, beer, considerably watered down, was a common beverage in medieval Europe; the alcohol content may have acted as a useful disinfectant for unreliable water supplies.[9]

Around eight thousand years ago, at the same time the Chinese were figuring out how to plant rice and get high by steeping it with honey and hawthorn fruit and letting it ferment, people in northern Europe came up with the brilliant idea of stealing milk from a cow's teat and chugging it themselves. A few years ago, I had a chance to give a presentation at Umeå University in northern Sweden and travel around the country for a week. I was impressed by the dairy-rich cuisine, redolent in cheeses and cream, and the attractive, statuesque Swedes. The link between height and dairy consumption has long been touted, so the size of milk-loving northern Europeans is not that surprising. More unexpected, however, is that milk-drinking nations have the world's highest rates of hip fractures.[10] Since we've all been taught that calcium is the basis for strong bones, the subject of milk and calcium generates a lot of head-scratching, angry tirades, and rebuttals. Is dairy healthy for humans?

The first thing to bear in mind is that milk is the most complicated substance that humans consume. Compared to milk, alcohol is child's play. Besides calcium, cow's milk contains phosphorous, saturated fat, casein and whey proteins, amino acids, the potent insulin-like growth factor 1 (IGF-1), antibacterial defenses (such as lactoferrin, lysozyme, and lactoperoxidase), immune system boosters (such as T cells, B cells, and immunoglobulin A), and a mad scientist's treasure trove of hormones, including gonadal hormones (estrogens, progesterone, and androgens), adrenal gland hormones, pituitary hormones (prolactin and growth hormone), hypothalamic hormones (gonadotropin-releasing hormone, luteinizing-hormone-releasing hormone, thyrotropin-releasing hormone, somatostatin), parathyroid hormone-related protein, insulin, calcitonin, and bombesin (influencing satiety, blood sugar, gut acidity, and gastrointestinal hormones).[11] New hormones in cow's milk are still being discovered, so the list will march on.

In the initial phase of experimentation, milk drinkers thousands of years ago must have suffered indigestion; however, if milk was consumed for a long enough time, adaptation in the colon took place, increasing fermentation of lactose and cutting back on the hydrogen gases. Once invented, products like cheese, butter, and yogurt lasted longer and contained less discomfiting lactose.[12] People in Africa got the same idea of using dairy, probably independently, because the genetic adaptations that allowed northern Europeans and East African pastoralists to digest lactose in milk involved different genes. Meanwhile, goat's milk was exploited in places like the Mediterranean and West Africa; Central Asian herders consumed horse milk (mare's milk has a protein and salt composition more similar to a woman's milk than cow's milk), and Bedouins did the same with their camels. (Bedouins could subsist almost entirely on camel's milk alone, a testament to the remarkable life-sustaining properties of milk.)[13] Historically, milk has also been obtained from sheep, water buffalo (milked in South and Southeast Asia; the source of authentic Italian mozzarella cheese), yak, and reindeer.[14]

The llama and its close cousin the alpaca seem to have been suitable candidates for dairying but were never milked in traditional times by their handlers in the Andean mountains (a region encompassing present-day Colombia, Ecuador, Peru, Bolivia, and Chile). Llamas were used to carry burdens, while the smaller alpaca furnished beautiful wool. As the geographer Daniel W. Gade has pointed out, the llama and alpaca are sufficiently docile animals, comparable to goats or sheep, that they could have been milked. The Inca were sophisticated enough to breed white llamas, but for some reason they never thought to tip back a cup of milk from their llama or alpaca. There is no trace of dairying in Incan history, art, or language. One possible reason llama and alpaca milking never took off is that dairying was a cultural invention that diffused across the Old World but was unable to penetrate into the Andean mountain ranges until the Spaniards arrived; the same holds true for other late cultural imports to the region, such as the wheel, the arch, and writing. Another factor that could have delayed the incorporation of dairy into the Andean cuisine is that locals already had calcium in their diets from the use of lime in potato sauces and as an accompaniment to coca chewing, and from the consumption of quinoa.[15]

Another potential barrier to the popularity of drinking milk is that it may strike the uninitiated as disgusting. Although the idea of drinking cow's milk is deeply rooted and celebrated in Western countries, plenty of Westerners today would hesitate before quaffing secretions from the teat of a horse, camel, or water buffalo, which are traditional drinks in other parts of the world. People in East Asia long disdained cow's milk as a barbarian concoction, and only with active government and industry intervention is the drink now making inroads there. If dog milk were demonstrated to be thoroughly nutritious, how many Western shoppers would toss a carton of it into their grocery basket? People develop an innate aversion to bodily fluids from another animal, because such fluids could bear a nasty infectious disease.[16] Thanks to their prominence in Western culture and society as sources of milk, the image of cows and their milk has been sanitized, but other potentially nutritious fluids, such as horse milk and pig's blood, strike us as unpalatable, unless we are acclimated to them from childhood.

Incidentally, the widespread consumption of dairy today may help to explain why acne is highly prevalent in dairy-consuming countries. A series of large-scale studies (47,355 participants in one study alone) conducted by the Harvard School of Public Health found an association between dairy intake and acne in teenagers. The reported prevalence of acne among adolescents in Westernized societies is between 79 percent and 95 percent, while traditional societies like the Kitavan Islanders in Papua New Guinea and hunter-gatherers in Paraguay have virtually no incidence of pimples. Some researchers contend that high-glycemic Western diets cause spiking in blood sugar levels and insulin, and hence unleash a cascade of androgen hormones and IGF-1, lowered sex-hormone-binding globulin, and increased activity in sebaceous glands, thereby exacerbating acne. Cow's milk is also known to raise IGF-1 levels among drinkers. However, skim milk seems to be a risk factor for acne rather than whole milk. Whole milk contains saturated fat, and researchers have discovered that saturated fats suppress bacterial activity, while monounsaturated fats (also found in vegetable oils and nuts) increase the incidence of acne, perhaps by increasing the permeability of the skin.[17]

Despite the popularly discussed link between milk and acne, many people may continue to drink milk for fear of having weak bones. The standard recommendation in America and Canada for adult calcium intake is 1,000 mg per day, compared to 800 mg in most of Europe and 500 mg in Japan. Who is right? As previously mentioned, in countries where people get more calcium in the diet, hip fractures are more common. Moreover, popping calcium supplements seems to increase the risk of hip fracture. Beyond a minimum threshold (around 400 mg of calcium per day), more calcium doesn't seem to help our bones. When I traveled in Papua New Guinea, there was no dairy intake at all among the villagers I stayed with, apart from mother's milk during infancy, yet their bodies were remarkably sturdy. New Guineans have among the lowest rates of hip fracture in the world. The highest rates of hip fracture occur among the statuesque milk-loving northern Europeans. Hip fractures are relatively easy to record and thus provide the clearest evidence with respect to calcium and bone health, but studies on osteoporosis generally show a disappointing lack of benefit with calcium supplementation, contrary to popular wisdom.[18]

(Soy products, on the other hand, are associated with decreased hip fracture rates among women, perhaps due to the effects of phytoestrogens, or the vitamin K found in fermented soy products such as stinky *natto* in Japan, and the even stinkier *doufu ru* in China and *tuong* in Vietnam. Fermented cheeses like aged goat cheese, blue cheese, brie, cheddar, and Parmesan are also rich in hip-preserving vitamin K, unlike nonfermented cheeses such as mozzarella and processed cheese.[19])

The role of dietary calcium, independent of milk intake, in promoting prostate cancer risk has been demonstrated in several studies. To consider one case, the Yoruba in Nigeria have no traditional history of dairying, and 99 percent of them are lactose intolerant. Most Yoruba have a gene variant that makes them more efficient at absorbing calcium (milk was absent in the traditional diet) and likely boosts their bone density. In a modern diet with a heavy daily dose of calcium, that super-efficient calcium absorption is a drawback, because it puts people with these genes at greater risk of developing prostate cancer. (Prostate gland tissues appear to be readily stimulated by calcium.) People of African descent are frequently carriers of this gene (71 percent of African Americans in the American Southwest, compared to 45 percent of Japanese in Tokyo and 20 percent of Utah residents of northwestern European ancestry). African Americans have a higher risk of developing prostate cancer, while the few African Americans who lack this gene appear to be at decreased risk, especially if they consume less calcium.[20]

Similar to the Yoruba, the Inuit also had a dairy-free, low-calcium diet. Inuit children consumed perhaps 20 mg of calcium per day from traditional foods. Since the Inuit are genetically adapted to low-calcium diets, Inuit children who eat a high-calcium Canadian-style diet often exhibit dangerous levels of calcium in the blood, a condition that can damage the kidneys.[21]

Conversely, in the case of pastoralists who consumed a lot of meat and especially milk, people developed genetic adaptations to deal with high loads of cholesterol from those foods. East African Maasai pastoralists became genetically adapted to a mega-cholesterol diet of cattle milk, blood, and meat, with two-thirds of their calories coming from fat alone. Their daily cholesterol intake was four to six times the average Western cholesterol consumption, but the level of cholesterol in the Maasai blood was far lower than Western levels. The genes of Maasai show evidence of modifications in regions related to cholesterol metabolism and synthesis, atherosclerosis (the thickening of arteries that is related to cholesterol deposits), and lactase persistence. All of these genetic adaptations seem to make the Maasai better suited to a milk-rich, cholesterol-heavy diet.[22]

In view of the traditional reliance of the English, Scandinavians, and northern Indians on milk products, it makes sense that most of them possess the enzyme lactase and are able to digest milk into adulthood (known as lactase persistence). Dairy use also left its genetic signature in widespread lactase genes among groups in East Africa and the Middle East, while in southern India and the eastern Mediterranean, the figure reaches around 15 percent. Few people were lactase persistent in West Africa, East Asia, and the New World.[23] Overall, two out of every three people in the world lack the ability to produce lactase.

In North America, a confluence of circumstances led to a boom in drinking milk. At the end of the nineteenth century, an abundance of pastureland close to cities plus improvements in milk storage techniques allowed production and consumption to soar. The U.S. Department of Agriculture (USDA), established in 1862 by President Abraham Lincoln, was charged with two mandates: to promote agricultural interests by increasing consumption of American agricultural products and to promote the health of Americans by setting dietary guidelines. The conflict of interest built into the USDA eventually materialized when the growing clout of the dairy industry led to the 1915 formation of the National Dairy Council, whose aim was to promote research on the benefits of dairy consumption. In 1919, facing a glut of milk supplies after the end of World War I, the USDA and the dairy industry began a program to increase milk drinking among schoolchildren. Educational materials featuring the merits of milk, including games and songs, were supplied with government endorsement. In Canada, meanwhile, the dairy industry succeeded in banning nondairy butter substitutes—namely margarine, originally made from beef fat—from Canadian markets in 1886. In 1948, the Supreme Court of Canada ruled that the ban was unconstitutional, after which provinces were free to set their own regulations on margarine production and importation. The last bastion of resistance was Quebec, which became the last place in the world to finally permit the sale of yellow margarine, in 2008.[24]

Although the interference of political and business groups in public health policies related to dairy consumption was regrettable, this interference does not directly address the question of whether dairy consumption is safe, and if so, for whom. As with all traditional food cultures, the cuisines in places in the world where dairy has a long history, such as northern Europe, pastoralist groups in East Africa, and northern India, performed well in meeting the nutritional requirements of eaters, in accordance with the kinds of animals and plants that could be raised and grown in these environments. Given their several thousands of years of exposure to dairy, over generations the people in these regions have evolved the genetic makeup to handle the challenges of digesting lactose, as well as other possible negative effects of dairy consumption. Conversely, in places that had little or no dairy consumption, such as the New World, people's traditional cuisines were adequate in meeting their nutritional requirements, including calcium intake, and so they may lack the genes to handle high loads of calcium, cholesterol, and other characteristics of dairy. In other regions, dairy products were a useful supplement to the diet, such as goat cheeses in the Mediterranean and ghee (clarified butter) in southern India, and should remain as useful food items. If we try to drastically alter traditional cuisines, either by adding a lot of dairy to dairy-less cuisines, or removing dairy from dairy-dependent cuisines, then we risk running into nutrient imbalances, because creating a tasty balanced diet from scratch is a lot tougher than eating something that was tested and savored over hundreds of generations.

One final aspect of dairy needs to be discussed. It is plausible that consuming a lot of dairy will increase a person's height, due to the presence of IGF-1 or other as-yet-undiscovered hormonal factors in milk. Height is highly socially desirable across societies, particularly for men, but being tall is also plausibly linked to a greater risk of certain forms of cancer, including prostate and breast cancers. In other words, copious dairy consumption involves a trade-off between health and height. This is not a trivial issue for parents to consider. Since height is a relative measure—for example, being of average height in Canada means being tall in Southeast Asia—the ideal outcome is that successive generations across the industrialized world gradually decrease in height to a level that is more optimum for long-term health; this way, no one's ego needs to be bruised for being small. On the whole, our species, at least in the industrialized regions of the world, has grown too large, beyond the scale that our bodies were built to handle. Small has to become the new beautiful.

# A TRUCE AMONG THIEVES

[I felt as if] my heart were suspended by a single thread. . . . My lips were observed to become pale. . . . A violent palpitation succeeded.

—J. RIDLEY, *"An Account of an Endemic Disease of Ceylon Entitled Berri Berri"*

One curious circumstance in connection with hay fever is that the persons who are most subjected to the action of pollen belong to a class which furnishes the fewest cases of the disorder, namely, the farming class.

—CHARLES BLACKLEY, *Experimental Researches on the Causes and Nature of Catarrhus Æstivus*

Ever since the importance of hunted animals in the diet began to decline, humans have sought suitable replacements. Rough kernels of wheat and rice, tiny cobs of corn, poisonous potatoes, and bitter olives were reborn as aromatic loaves of bread, elegant ribbons of pasta and noodles, bowls of fluffy rice, filling tortillas, hearty potato stew, and nourishing olive oil. The milk issuing from goat, sheep, cattle, camels, and horses was ingeniously fashioned into yogurt, cheese, and butter. Domesticated animals provided much-appreciated meat when they were no longer able to earn their keep through pooping, peeing, tilling, carrying heavy loads, assisting in hunts, laying eggs, or killing rodents and other agricultural pests.[1] We tamed wild plants and animals; we multiplied our numbers. Forests and riverbanks that once harbored rich ecosystems were taken over by villages that mushroomed into towns. People could now spend their days indoors, shielded from the harsh elements, the violence inflicted by marauders, and the boredom of talking to the same villagers night after night. The new urbanites traded their expertise in the manufacture of goods and the provision of services for currency that could be redeemed for housing and social stimulation, as well as food that was hauled in from the peripheries of towns.

Although many changes in diet and lifestyle were beneficial on the surface, rapid changes sometimes carried grave unintended consequences. As we have seen, our ancestors had been undergoing changes in diet and environment for millions of years. Our diets shifted from insects to fruits, meats, and agricultural products like wheat, rice, potatoes, and corn, then added milk and alcohol. By contrast, viewed against this backdrop of gradual transformation, the last thousand years of human history has been a storm of disruption due to technological and scientific breakthroughs. Since biological evolution requires dozens or hundreds of generations to adapt organisms to new environments and foods, new afflictions began to appear. Some of these could be contained and defeated, but other outbreaks took their place, gathered force, and tore through populations with brutal ferocity.

As recounted in an excellent study by Kenneth John Carpenter, beginning in the seventh century, observers in densely populated areas of East and Southeast Asia periodically described an ominous constellation of symptoms including tremors, numbness, difficulty in walking, swelling of the limbs, wasting, and pervasive weakness; when the heart began to race, death was usually not far away. The patterns of the disease befuddled observers. It was prevalent in southern China but not northern China. Unlike cholera, which vaulted into Japan via seaports linked to China, this disease was not contagious; people who moved from afflicted regions did not carry the disease with them.

Japanese doctors tried acupuncture and raising blisters along the spine with heated cylinders but to no avail. Western doctors theorized that noxious airs were the root cause of the affliction, but the disease was common on ships manned by Asian crews. In the 1870s, the rage in medicine was bacteria. Louis Pasteur successfully treated cholera and anthrax through manipulating bacteria; Robert Koch found the bacterial agent responsible for tuberculosis. Perhaps the disease afflicting East and Southeast Asia was another bacterial epidemic? However, experiments on chickens failed to reveal any bacterial agent.

People who ate barley did not seem to be affected, which focused attention on the possible role of a diet of rice. This led to another puzzle: Poor-quality rice did not seem to increase the risk of acquiring the disease; if anything, those who were privileged to eat better-tasting rice were more susceptible. Adding to the confusion, the disease, known in the Dutch Indies as beriberi, would also appear in areas of the world where rice was not eaten, such as Brazil and Canada.

Two surgeons, Japanese and Dutch, working with different navies, found that the disease could be greatly alleviated by adding sources of protein to the diet.[2] Although this was a relief for the navy men, rats fed on protein-adequate but otherwise nutritionally deficient diets failed to thrive, which disproved the protein hypothesis. Further experiments revealed that apart from protein, fat, and carbohydrates, rats required two additional substances for survival: a fat-soluble "vitamin A," which could be obtained from cod liver oil and butter, and a "vitamin B," which could be obtained from yeast, wheat germ, and nonfat milk powder. Further work isolated a vitamin $B_2$ complex, and a vitamin $B_1$ later named thiamine, which was able to prevent beriberi-like symptoms from appearing in rats and chickens. Purified crystals of thiamine were effective at restoring the health of rats and chickens even when administered in microscopic quantities.

Thiamine, it turns out, is present in the bran of rice and is removed when rice is milled, heated to very high temperatures, or boiled and then rinsed. Milling boosted the storage life of rice and increased its palatability but inadvertently removed thiamine in the rice germ from the diet. By introducing steam-powered milling to Asia, the colonial powers greatly increased the misery wrought by beriberi, though the Chinese and Japanese also used rice mills and were beset with the disease. (Diets of cassava in Brazil and bread made from white flour and baking powder in isolated ports of Newfoundland also lacked thiamine and hence led to outbreaks of beriberi in those regions.)

There were two relatively simple ways of resolving the thiamine deficiency. The first method was to parboil rice, a traditional way of cooking rice in parts of South Asia that involves soaking it and then boiling it in its husk. This aids in removal of the husk but also helps the germ to retain nutrients from the husk, including thiamine. During the epidemic of beriberi in Asia, populations that used parboiled rice were not affected. However, parboiled rice had a musty smell and a yellow-brown tinge and was not as fluffy as white rice, which made it unacceptable to East Asian populations.

The second traditional way of preparing rice was to pound or stamp upon rice kernels, then employ sifting or other means to remove the husks. Because this method only incompletely removes the husk, the rice still retains thiamine in the "silver skin" surrounding the kernel. However, people used to eating white rice also rejected hand-milled rice as unpalatable, defeating the efforts of public health officials. A third method was to combine thiamine-rich beans with white rice, a practice still carried on in parts of Asia. Eventually, the disease was resolved definitively by adding thiamine directly to polished rice, but not before beriberi had inflicted much misery upon populations.[3]

While beriberi was ravaging urban areas in East Asia, physicians in Europe were coming across a novel disease whose symptoms consisted of blisters, loss of appetite, depression, and a preoccupation with suicide. The disease was dubbed "pellagra," a term denoting rough skin in the Lombard dialect of Italian. Unlike beriberi, pellagra had a tendency to afflict the poor rather than the rich, the opposite of beriberi's pattern. The locations of the disease were also opposite, with most cases of pellagra being recorded in Europe, while beriberi primarily afflicted East and Southeast Asia.

American doctors probably spotted the first cases of pellagra in the nineteenth century, but since the disease was believed to be nonexistent on their side of the Atlantic, they refrained from announcing their observations. In 1902, a physician in Atlanta identified the disease in an impoverished farmer. The visibility of pellagra rapidly expanded. In 1906, there were eighty-eight cases of pellagra at Mount Vernon Hospital for the Colored Insane in Alabama. Eighty of these patients were female, and more than half died. Mysteriously, none of the nurses at the hospital contracted the disease. Other mental institutions reported outbreaks, and the epidemic spread as far west as Illinois. By 1912, around twenty-five thousand cases had been diagnosed, with four in ten victims dying. As with beriberi, expert opinion initially focused on microbial agents. Some people believed that pellagra arose from eating spoiled, moldy corn, and thus several states enacted laws to inspect corn. Pellagra was also thought to be infectious, and so pellagra victims, who invariably came from the most economically disadvantaged quarters, were shunned like lepers and denied access to hospitals.[4]

In February 1914, the U.S. surgeon general invited a talented Jewish Hungarian American epidemiologist, Dr. Joseph Goldberger, to take over the Public Health Service's faltering pellagra investigations. By the time of his appointment, when he was forty years old, he had made a name for himself studying—and surviving—epidemic diseases. Shortly after launching his investigations, Dr. Goldberger surmised that the disease was not communicable, since health workers who were in close association with pellagra victims never acquired the disease. A more likely explanation lay in the classic "Three M's" diet of poor southerners: meat (fatty pork), molasses, and meal (cornmeal). Orphans and mental institution patients who ate monotonous meals along the Three M's pattern got pellagra, but workers at the same institutions who had access to more varied fare avoided the disease.

Dr. Goldberger carried out an experiment with volunteers at a

Mississippi prison, who were offered pardon by the governor as a condition for their participation. Over the course of six months, more than half the volunteers who were fed a diet based on cornbread and cornstarch developed skin lesions (starting with the genitalia), while the remaining subjects developed less striking but still noticeable manifestations of the same disease. Although the experiment was carried out with great meticulousness by Dr. Goldberger, both he and the Mississippi governor were roundly criticized for its unorthodox nature. Moreover, not only did the results run contrary to the conventional line that pellagra was infectious, the nutritional hypothesis also drew attention to the poverty of the South, which provoked the ire of proud southern politicians and patriots.[5]

Dr. Goldberger continued to try to convince critics that pellagra was noncommunicable, even going to the extraordinary extent of injecting himself, his wife, and colleagues with blood from pellagra victims, and swallowing skin scales, feces, and dried urine from pellagra sufferers, wrapped up in dough. Consuming this concoction yielded nausea and diarrhea, but no pellagra. Unable to sway his critics, but convinced by his own observations and efforts that amino acid deficiency rather than spoiled corn was key, Dr. Goldberger then tried to locate the missing amino acid. He died of renal cancer in 1929 before he could complete his life's mission.

In the end, it turned out that his hypothesis was correct: Corn was found to be deficient in tryptophan, which the human body can metabolize into niacin (also known as vitamin $B_3$). By the 1940s, fortification of foods with vitamin $B_3$ eliminated pellagra as a menace to poor Americans, though not before some 3 million had suffered from the disease, resulting in approximately 100,000 deaths. In Italy, pellagra cases peaked in the late nineteenth century among poor rural peasants in the north confined to monotonous diets of corn, then faded as economic conditions improved through emigration (which raised local wages and brought remittances from emigrant workers), industrialization, improvement of crop yields, and falling prices for wheat (which was substituted for niacin-deficient corn). Pellagra disappeared from Italy by the 1930s.[6]

Industrial milling of American corn began in the early 1900s, which stripped the corn germ and thereby boosted the shelf life of processed corn. Unfortunately, the germ is also where niacin resides. During the epidemic, rates of pellagra were highest in areas adjacent to railroads, where people had ready access to stores and industrially milled cornmeal. In rural areas, people relied instead on traditional processing techniques, such as water-driven stone-milling, which preserved more of the corn germ and reduced the risk of pellagra. Indigenous groups in the Americas who domesticated corn over hundreds to thousands of years knew how to prepare their sacred crop for safe consumption. Through trial and error, and copying neighbors, tribes that relied heavily on corn learned to cook it with an alkaline substance such as lime or wood ashes, which helped to increase the availability of tryptophan and niacin in corn and thus helped them avoid pellagra. Another method of preserving niacin, practiced by the Tohono O'odham and the Hopi Indians, was to roast immature corn, which contains higher concentrations of niacin than mature corn.[7]

While beriberi was crippling swaths of East Asia and pellagra was picking up steam in southern Europe, a different disease was ravaging northern European cities. In 1634, fourteen deaths in England were attributed to a condition in which children were left with a deformed spine and chest and crooked arms and legs. The disease had shown up in the Balkans in 9000 BC, in early Egypt, and in China around 300 BC, but it finally developed into a full-scale epidemic in industrialized urban European locales in the eighteenth century. Nor was this just a disease of children, as elderly women in northern European and North American cities and towns suffered from high rates of bone fractures.[8]

Since prevailing medical theory centered on "humours," the condition known as rickets was blamed on cold distemper. Herring, a rich potential source of vitamin D, was banned as "cold" food. Peasants found their own cure for the disease by consuming raven livers (livers are key organs in vitamin D metabolism). Fishermen around northern Europe had been taking fish liver as a household remedy for centuries. Swallowing cod liver oil was another matter, for it was prepared by allowing livers to spoil until the oil could be skimmed from the surface. The stench, understandably, was nauseating.[9]

As medical practitioners continued to debate the merits of cod liver oil, sunlight, bloodletting, bone breaking and resetting, and racks and slings designed to stretch out children's bodies, rickets accompanied settlers migrating to the New World. Between 1910 and 1961, 13,807 deaths in the United States were officially attributed to rickets, mostly in infants less than a year old.[10] Dark-skinned children were especially vulnerable, particularly in northern cities. Finally, between 1919 and 1922, a series of experiments conducted by researchers in Vienna verified the efficacy of cod liver oil and sunlight in preventing and treating rickets, and thereafter supplementation with cod liver oil, vitamin-D-fortified milk, and the use of sunlight gradually eradicated rickets. However, cases continue to occur even to the present day.

Although the scourges of beriberi, pellagra, and rickets are largely behind us, there are important lessons to learn from the history of these diseases. In each of these cases, a major rethinking—a paradigm shift—was required before progress could be made. In the case of rickets, the old theory of humors made European medical experts skeptical that herring, a "cold" food, could be beneficial, even though we now recognize that herring is a rich source of vitamin D and would have helped to alleviate rickets—a much better cure than the racks that were used to stretch out the deformed bodies of children afflicted with the condition. In the case of beriberi and pellagra, medical opinion clung to the notion that these diseases were caused by infectious germs, thus adding crucial delay to the search for nutritional deficiencies.

Once the role of vitamins was understood, progress was rapid. Adding vitamins like $B_1$ (thiamine), $B_3$ (niacin), and D to factory-produced foods was straightforward; such measures required no alterations in habits, and because the vitamins could be produced cheaply, no one protested the additions. Moreover, the companies that produced the vitamins profited handsomely, and thus the disasters of beriberi, pellagra, and rickets were averted in the ways in which capitalistic societies operate most comfortably: with the scent of profit, the comfort of cheap goods, and minimal prodding from public authorities. Unfortunately, the scent of profit and the lure of quick vitamin fixes continues to dazzle the public: The U.S. supplement industry registered a strapping $28 billion in sales in 2010 despite a lack of evidence for benefits from taking vitamins and antioxidant supplements among today's well-nourished population.[11] The lure of "superfoods" is similarly dubious.

Today, industrialized societies are facing several new epidemic diseases that are being studied for potential treatments; however, because the basis of these diseases conflicts strongly with past medical understanding of how the body works, there is a lot of conflict and confusion among medical experts and the public. We'll consider two paradigm-breaking clusters of diseases: sunlight-deprivation diseases and allergic diseases. It now appears that myopia (i.e., nearsightedness) and allergic diseases are triggered by radical lifestyle shifts that were undertaken in recent centuries and decades. Understanding myopia forces us to reconsider the role of sunlight in guiding the development of the eye, and reining in allergic diseases compels fresh thinking on hygiene and bacterial warfare, as well as the influence of sunlight and vitamin D on the immune system.

From the point of view of evolution, nearsightedness is a great mystery. Back in hunter-gatherer times, anyone unable to spy a stalking predator or a tasty morsel in the forest would have been at a tremendous disadvantage. Myopia was first described by the ancient Greeks, but in the two thousand years since, no one has ever come up with a good explanation for why myopia developed in some people and not others. The old theory was that engaging in too many near-work activities, like reading or writing (or these days, using a computer or smartphone or playing with handheld video games), resulted in prolonged tension of eye muscles and eventually permanent myopia. This theory was proposed at least as far back as 1866 and seemed to make sense, since children first develop myopia during the early school years and myopia is more common among white-collar occupations and rises with education level. However, empirical studies show mixed results about the alleged effect of near work on vision, and the use of various types of lenses to correct for near-work effects have so far been unable to halt myopia from progressing in children. Meanwhile, the prevalence of myopia is rising in regions like East Asia. For example, in Singapore, the prevalence of myopia nearly doubled over a two-decade span, reaching 43 percent among young men.[12]

In striking contrast to the muddled empirical results of the near-work theory of myopia, in back-to-back studies in three different countries, children who play outside more frequently were found to be less nearsighted (Australia, the United States, and Singapore).[13] The most solid explanation for this pattern is that sunlight is protective against nearsightedness. This pattern has been replicated in controlled experiments with chickens and monkeys, and also in a study that looked specifically at ultraviolet exposure and myopia. The reason sunlight helps prevent myopia could be the greater depth of focus and clearer retinal image achieved in bright sunlight or the stimulation of dopamine from the retina by sunlight. Sunlight's protective effect may help explain why myopia rates are lower in Europe than in East Asia: Blue eyes have very little melanin in the iris compared to brown eyes and hence may permit greater intensity or different wavelengths of light to impact the pupil. Additional studies will be required to develop a complete account of the mechanics underlying myopia, but in the meantime, some people will find ways of increasing lighting inside homes so that it more closely mimics the tremendous intensity of natural sunlight; conversely, they could opt to let their kids play outside more.[14]

In addition to lowering the risk of nearsightedness, bright light triggers serotonin production in the human brain and combats the misery of seasonal affective disorder (SAD) and depression. Among patients admitted for treatment of depression at a Canadian psychiatric ward, those who by chance received one of the sunny east-facing rooms checked out of the hospital nearly three days sooner than those who were allotted one of the dimmer rooms. The antidepressive effects of sunlight may go beyond shortening hospital stays: Among patients admitted to the cardiac intensive care unit for heart attacks, those who stayed in one of the dim rooms had a higher chance of dying than those who received one of the sunny rooms. Over a four-year stretch, 13.2 percent of the patients who stayed in one of the four dim north-facing rooms died, compared to 7.7 percent of those who received one of the four sunny south-facing rooms.[15]

Sunlight or geographical latitude effects have also been noted in the incidence of schizophrenia and autism, ailments that have befuddled researchers up until now. Colder northern countries have the heaviest burden of each of these diseases (and in the case of schizophrenia, children of darker-skinned immigrants are particularly susceptible), leading researchers to investigate whether melatonin disruption, vitamin D deficiency, or some other factor is causing the association of these diseases with paucity of sunlight.[16]

Left to its own devices, the skin regulates production of vitamin D from ultraviolet light (UVB, to be specific) to manageable levels, just as our body does with all our other hormones. However, there are two major problems with relying solely on our skin to provide vitamin D. First, there's the problem of our beautiful birthday suits. Human skin pigmentation evolved over thousands of years to provide the optimal balance of vitamin D production, protection against cancer-inducing ultraviolet light, and protection against damage to folate levels (folate, or vitamin $B_9$, is easily damaged by ultraviolet radiation). When humans migrated out of Africa into Europe and East Asia, skin types in these latter two regions independently evolved to become lighter, powerful evidence that sunlight was an important factor influencing mortality. However, you can't change skin color like a coat, so when Europeans started to populate sunny colonies in the Americas and Oceania beginning a few hundred years ago, and people from the tropics, like my parents, moved in the opposite direction, to frigid climes, the wonderfully adapted skin color suddenly became a liability. My Caucasian friends get sunburn from the Californian, Australian, and Southeast Asian sun, while my tropical immigrant friends and I languish from sunlight deficiency in northern cities like Ottawa, Umeå, and Sapporo.[17]

The second problem with relying on skin for our vitamin D is our pattern of sun exposure. Some people are able to tan, which is the skin's method of adapting to the rise and fall of ultraviolet rays over the seasons. These days, when sun-starved office workers dash outside to play on weekends, then spend the rest of the week working inside, the alternation between scorching and seclusion sets us up for increased risk of sunburn and developing cutaneous melanoma, the most aggressive form of skin cancer. To avoid skin cancer, people slap on sunscreen, but it's unclear whether this practice helps or harms, since sunscreen may give people a false sense of security and encourage them to spend more time outside, and the pattern of sunscreen wearing off and then being reapplied may exacerbate the dangerous intermittency of sun exposure. On top of that, the depletion of the ozone layer may be increasing our exposure to ultraviolet radiation beyond the range that our skin is adapted to handle. (Ozone pollution in big cities may cause the opposite effect and screen out UV light from reaching us.)

Not surprisingly, many people, for cultural or health reasons, decide to forgo the hazards of ultraviolet radiation altogether by seeking refuge under parasols, long-sleeved clothing, or heavy sunscreen use, but then they suffer from vitamin D deficiency, and we're back to square one. Others worry about vitamin D deficiency and pop vitamin D pills, but the problem is that no one knows exactly how much vitamin D is a healthy dosage or how vitamin D supplements influence our immune system and increase our risk for diseases like cancer. In my own case, I love Canada, especially during the quiet, languid summers, but the mismatch between my skinny brown body and the rigors of a Canadian winter is so uncomfortable that I spend as much time in the tropics as my schedule and meager budget allow.[18]

Most allergic diseases were unheard of only decades ago. In the West, there were two waves of allergies. Asthma initially surfaced fifty years ago and reached its highest rate in the 2000s. Hot on its heels, food allergies have besieged Western countries. In one of the most ambitious screening tests for food allergies yet conducted, more than 10 percent of infants in Melbourne were shown to have a food allergy to either peanuts, eggs, or sesame seeds, a higher rate than pediatricians and scientists had previously suspected.[19] In Asia, rates of asthma are now rapidly rising, and it is expected that food allergies and eczema will arrive soon afterward. Curiously, some kinds of foods that trigger allergic reactions in Asia are turning out to be rather novel. For instance, in Singapore, the most common trigger of anaphylaxis (the rapid onset of severe allergic symptoms, including hives and difficulty in breathing) is edible bird's nests, a Chinese delicacy made from the saliva of cave swifts. Buckwheat is a common allergen in Japan, while the same is true of chestnuts in South Korea and chickpeas in India. As in the West, eggs, milk, and sesame are also commonly reported to induce allergic symptoms.[20]

Why is the allergy epidemic rearing its ugly head now, and what can be done about it? Scientists have analyzed three large-scale changes in traditional diets and lifestyles that are suspects in the allergy epidemic. The first is the shift in the human diet in industrialized countries from inflammation-calming omega-3 fatty acids to inflammation-inciting omega-6 fatty acids. This dietary transformation resulted from feeding livestock with seeds like corn, instead of allowing them to indulge in their natural diets of grasses or insects, from the widespread use of processed seed and vegetable cooking oils, and from reliance on industrially produced foods instead of harvesting wild plants and animals. A number of recent studies have scrutinized the relationship between childhood allergic diseases and consumption of omega-3 and omega-6 fatty acids. Babies born to mothers who took fish-oil supplements show fewer allergic symptoms to cat allergen and egg than babies born to mothers who took olive oil supplements.[21] Children who eat more fish in early life have lower risks of asthma, eczema, allergic rhinitis (commonly called hay fever, though allergic rhinitis includes allergies to more than just hay), and produce fewer antibodies in an allergen blood test.[22] Children whose mothers had a history of asthma but who ate oily fish at least once a month during pregnancy had a lower incidence of asthma. On the other hand, if their mothers were fond of eating fish sticks during pregnancy, the kids had a higher risk of developing asthma. Fish sticks are made from cod or pollock, which have low levels of omega-3 fatty acids. Moreover, American fish sticks are only 40 percent to 72 percent fish flesh by weight, and the batter coating is made from fried corn, canola, cottonseed, and soybean oil, which are heavy in inflammatory omega-6 oils. To make matters worse, when these oils are heated to high temperatures, they transform into trans-fatty acids, which are notorious for increasing inflammation.[23] Much the same result was obtained in a German study: If mothers ate margarine or vegetable oil (both high in inflammatory omega-6 fatty acids) during their final month of pregnancy, their children were more likely to develop eczema by two years of age. If mothers ate fish in the final month of pregnancy, their children were less likely to develop eczema, as one would expect if omega-3 fatty acids truly help to reduce allergic diseases.[24]

While the logical arrows and evidence pointing from inflammatory omega-6 and transfat-rich foods to allergies seem straightforward, so far omega-3-rich diets have not been proven to be a magic wand for banishing allergic symptoms, at least for adults. A reasonable conclusion is that what your mother ate or what you ate as a child may be the key, rather than your best efforts later on in life, at least as far as omega-3/omega-6 modification is concerned. Pediatricians currently recommend mothers breastfeed up to four or six months of age to minimize the risk of infants developing allergic disease. Good to know, but not much help for the average Joe or Jane dreading the arrival of hay fever season.[25]

A second major set of theories about the allergy epidemic centers on the role of vitamin D. This interest developed in part because vitamin D receptors are found in nearly all immune system cells and because many immune system diseases have been observed to increase the farther one gets from the equator. Vitamin D deficiency has been studied in connection with asthma, allergic rhinitis, food allergies, and eczema.[26] Asthma afflicts around 300 million people worldwide. Higher levels of vitamin D in the blood of children have been associated with lower rates, or better control, of asthma. Mothers with higher intake levels of vitamin D give birth to children who have lower rates of wheezing (a symptom often associated with asthma).[27]

Finland is a good place to check for a link between vitamin D deficiency and allergic rhinitis, because hours of sunlight there are scant and asthma rates are high. Researchers asked Finnish mothers who had given birth what they ate during their final month of pregnancy. When mothers acquired more vitamin D from food and supplements during their last month of pregnancy, their children had a lower chance of developing either allergic rhinitis or asthma.[28] This heightened impact of vitamin D on allergic rhinitis during a critical developmental window echoes the timing of the intake of omega fatty acids and the manifestation of allergic diseases.[29]

Is the risk for developing food allergies also heightened by lack of vitamin D? Until recently, no one had a clue, in part because food allergies are defined differently in different places and studies of food allergies tended to focus on small populations. A breakthrough came in 2002, when researchers in Manitoba hit upon the idea of using EpiPen prescription data to study anaphylaxis rates. EpiPens are self-injectable devices that deliver an emergency dose of epinephrine (also known as adrenaline) in case of an allergy attack, the most common being food-induced anaphylaxis. In Canada and other countries, prescriptions are required to obtain EpiPens and related devices. Thus data on EpiPens gave researchers a detailed view for the first time on patterns of anaphylaxis. When doctors scoured American data on EpiPen prescriptions, they discovered that northeastern states had the highest rates of epinephrine prescriptions, while southwestern states had the lowest rates. The leader in anaphylaxis misery was snowy Massachusetts. The lucky losers in this competition? Who else but Hawaiians, with New Mexicans and Californians close behind.[30] The difference in EpiPen prescriptions between New England and other American regions remained true after controlling for possible differences between states in terms of factors such as age, sex, race, income, health insurance, and the number of allergists, pediatricians, adult primary care providers, and emergency physicians. A similar pattern emerged in Australia, where EpiPen prescriptions and anaphylaxis admissions among children were more frequent in colder states such as Tasmania.[31]

Recently, scientists and doctors in the United States and Australia have announced findings that provide additional support for the vitamin D/food allergy hypothesis. Emergency department admissions for acute allergic reactions are more common in the frigid American Northeast than in the Southwest, and kids under the age of five living in Boston who were born in the fall or winter season, stretching from September to February (and therefore likely to be deficient in sunlight and vitamin D), are more likely to develop food allergies, particularly to peanuts. Gloomier Australian states have higher rates of prescriptions for hypoallergenic baby formula than sunny states, and kids living in those less-sunny states are also more likely to develop peanut and egg allergies (as well as allergic eczema).[32] Bottom line: Being born or growing up in a cold climate probably puts you at greater risk of developing asthma, food allergies, and eczema, particularly if you are darker-skinned or spend a lot of time indoors.[33]

Climate therapy targeting eczema since the 1950s has seen well-heeled Europeans flocking to the Baltic Sea, the French Atlantic and Mediterranean coasts, resorts in the Canary Islands, and other temperate areas around Eastern Europe and Western Asia for relief. Though the efficacy of sunlight on eczema was undeniable, only in the past decade have scientific trials been carried out to test this effect. One of the pioneering studies was done in Boston, where kids who took vitamin D daily showed improvement in their eczema symptoms. (This was a very small study involving eleven children.) Researchers in Iran reported similar success with vitamin D treatment of eczema. In a study of Italian children with eczema, the severity of their symptoms varied in direct proportion to vitamin D levels.[34] In 2008, researchers at the University of California, San Diego, noted that people who had eczema produced more antimicrobial amino-acid chains known as cathelicidin in the diseased portions of their skin; taking vitamin D supplements greatly elevated production of protective cathelicidin amino-acid chains.

In the last seven years, scientists cracked the genes that underlie one kind of eczema, ichthyosis vulgaris ("fish-skin disease"). Around 9 percent of people of European ancestry, 7 percent of Singaporean Chinese, and 4 percent of Japanese carry a genetic mutation that prevents their skin from manufacturing properly functioning filaggrin (filament-aggregating protein), a protein that normally teams up with lipids to keep water, microbes, irritants, and allergens out of the body. Without filaggrin, the skin becomes dry, scaly, and itchy, and a person becomes more susceptible to allergic reactions through the impaired skin. No one has yet offered a good reason why genes that predispose people to eczema are so widespread. It has been suggested that a more permeable skin in people with only one copy of the mutated filaggrin gene could have enabled "natural vaccination" through enhanced exposure to low concentrations of infectious diseases like tuberculosis and influenza; two copies of the same gene cause severe eczema and would have been a disadvantage, but the mutated filaggrin genes persist in populations because of the advantage gained by people who only have one copy of this gene.

Alternatively, since eczema often disappears in hot humid weather, eczema may only be a factor in contemporary industrial societies where people spend a lot of time in buildings that are equipped with heating or cooling systems that produce dry air, and where hot showers that strip away the skin's oils are considered a mandatory part of everyday life. My mother and I were able to tolerate torrid summer heat more easily than the rest of our family, and we both disliked air-conditioning. There are many more variants of mutated filaggrin genes in hot humid Asia compared to European populations, supporting the idea that there has been longer evolutionary selection for more permeable skin in hot humid climates.[35]

Our view of the effects of sunlight on health is rapidly expanding. This is both good and bad news. The good news is that our bodies are designed to tolerate and utilize a lot of sunlight, and therefore the details behind vitamin D, sunlight, and health are interesting but not crucial, as long as we receive adequate sunlight. The bad news, for people who avoid sunlight, or live in much hotter or colder regions of the world than their ancestors did, or work exclusively indoors, is that there is no quick fix for sunlight deprivation, such as taking vitamin D supplements. Although growing evidence seems to implicate vitamin D deficiency in causing allergic diseases, in a few studies, scientists have observed that more vitamin D means a *greater risk* of getting an allergic disease. For example, researchers analyzing survey data on 18,244 men and women across the United States found that white Americans who had higher levels of vitamin D were more likely to have been diagnosed with allergic rhinitis. Meanwhile, researchers poring over Finnish health records spanning three decades concluded that people who had received regular vitamin D supplementation (greater than 2,000 IUs, or international units, per day) in their first year of life had a greater likelihood of developing allergic sensitivities, allergic rhinitis, or asthma as adults. In Sweden, children who had more vitamin D from food and supplements were more likely to develop eczema. In England, mothers who had high levels of vitamin D in blood tests were more likely to give birth to children who developed eczema and asthma. Even more worrisome, researchers are observing that high levels of vitamin D seem to increase risks of esophageal, pancreatic, and prostate cancer in men. Frank Garland, one of the founding investigators and prominent boosters of vitamin D, died of esophageal cancer at age sixty.[36]

Some vitamin D supporters have criticized these studies for weaknesses such as small sample sizes and failure to control for confounding variables such as additional items in the diet outside of the study period. Other researchers suggest that the real danger lies in excessively high dosages of vitamin D. An additional consideration in taking vitamin D supplements is that vitamin D is a hormone. Hormones control the timing of our physical development, behavior, and reproduction, and thus do not take kindly to efforts by humans to meddle with their operations. Take, for instance, the roller-coaster history of postmenopausal hormone replacement therapy. Lured by promises of relief of symptoms of menopause such as hot flashes and vaginal dry-

ness, purported lowered risks of cardiovascular disease and hip fracture, and clever pharmaceutical innovation (combining estrogen plus
progestin to reduce the risk of endometrial cancer), 90 million women
in the United States took oral, transdermal, vaginal, and injectable
estrogens in 1999, the heyday of postmenopausal therapy. Then,
starting in mid-2002, a series of reports highlighted hazards such as
increased risk of breast cancer and cardiovascular disease, causing
a mass exodus from estrogen therapy and likely triggering a one-
time 6.7 percent drop in U.S. breast cancer rates in 2003. Menopause is a biologically programmed cessation in reproduction, which
seems to have evolved to spare ancestral women the danger of giving birth at a period when their existing offspring or grand-offspring
were still highly dependent on them for survival (the hazard of child-
birth is compounded in humans due to the constraints of narrow
hips adapted for walking upright and the relatively large heads of
human babies). We cannot easily or safely override this evolutionary imperative. The same can be said of other attempts at hormone
manipulation, such as the use of human growth hormone to reverse
aging in the elderly or doping with anabolic steroids for enhanced
athleticism, both of which are now understood to carry substantial
long-term health risks that outweigh their benefits.[37] Our bodies evolved
to regulate vitamin D use through exposure to sunlight rather than
vitamin supplements; circumventing this system by taking pills carries risks. The interaction of vitamin D with our bodies is too complex for scientists to guess at the proper dosage. This situation arises
time and time again in the history of nutritional supplements. Hopes
were raised and then dashed for beta-carotene, vitamin A, and vitamin
E supplements, among others. This type of conflicting advice seems
discouraging until one realizes that maintaining good health con-
sists primarily of finding foods and lifestyles that reflect the conditions of our ancestors and then letting our bodies—exquisite products
of millions of years of evolutionary refinement—do the rest of the
work.

So far, we have discussed two of the three major hypothesized reasons for the increase in allergic diseases, omega-fatty-acids imbalance, and vitamin D deficiency. In the nineteenth century, observers in England noted that people who suffered from hay fever, or allergic rhinitis, came disproportionately from the upper classes, which raised the possibility that something about education or race was involved in the condition.[38] In 1966, researchers in Israel observed that the incidence of multiple sclerosis *increased* with better sanitation, such as cleaner drinking water, less crowding, and the availability of flush toilets. A major breakthrough came in 1989, when Professor David Strachan, a lecturer in epidemiology at the London School of Hygiene and Tropical Medicine, observed that children with older siblings in the same household had lower rates of hay fever and eczema. Now, if a researcher looks at a large database with many cases and many factors, some of those factors will coincide purely through chance, just as cloud or rock formations may resemble a dragon or horse. What made Strachan's study striking was that for each additional older sibling, there was a corresponding drop in the likelihood that a child developed hay fever or asthma, as if having siblings were medicine whose effectiveness increased with each older brother or sister. Such regularity in a trend is highly unlikely to arise by chance. To top it off, Strachan had an insightful suggestion for why this pattern existed: When children come down with illness, they get their siblings (and parents) sick as well. A history of infectious disease was somehow protecting children against developing allergic diseases in later life.[39]

The "hygiene hypothesis" sparked resurgent interest in the connection between infectious and allergic disease. Scientists discovered that children showed improvement in symptoms or markers of allergic diseases if they:

- were not born via Cesarean section;
- owned a furry pet;
- attended day care;
- lived in a crowded household;

- were less frequently bathed or had their hands washed less frequently;
- were not vaccinated;
- received fewer antibiotics early in life;
- were exposed to bacterial toxins in mattresses;
- lived on farms;
- were exposed to farm animals;
- drank unpasteurized milk;
- had no access to sewage or clean water;
- were infected by viral diseases such as hepatitis A, herpes, or measles;
- were infected by noxious bacteria such as salmonella, *Helicobacter pylori* (induces inflammation of the stomach lining), or *Mycobacterium tuberculosis* (responsible for tuberculosis); or
- were infected by parasites or parasitic diseases such as malaria, giant roundworms, hookworms, flukes, pinworms, whipworms, or *Toxoplasma gondii* (usually infects cats, but can also be carried by rats and humans).[40]

As with omega-3 and vitamin D, the earlier in life exposure to pathogens and toxins occurs, the greater the reduction in allergic disease tends to be. The first item on this list and possibly the second (furry pets and day care) are likely the only ones that many Western children have encountered. Other exposures help protect farm children from allergic diseases. The last four items will send a shiver down the spines of Western parents, but such conditions are the lot of most kids in the developing world and indeed of most humans throughout history until recently.

As Ethne Barnes has catalogued in a marvelous book, *Diseases and Human Evolution*, the invention of agriculture was a godsend for microbes. Your average bacterium, virus, protozoan, or other parasite wants nothing more than affordable housing, opportunities for romance and sex, and plentiful food. To be sure, prior to the Agricultural Revolution, the life of hunter-gatherers had been no disease-free picnic. As previously mentioned, a diet of raw or undercooked meat was a good way for our ancestors to pick up a friendly tapeworm, which, although it conveniently prevented other tapeworms from colonizing the gut, did have the worrying ability to grow up to fifty feet long. Fortunately, due to their long evolutionary history with us and other mammals, tapeworms rarely cause significant health problems and live in mostly peaceful relationships with their hosts.[41] On the other hand, the larvae of the parasitic worm *Trichinella spiralis* (which causes trichinosis) are passed on when their hosts are eaten raw or undercooked by carnivorous animals, and they don't mind surging into muscle tissues to overwhelm host immune systems, thereby crippling their hosts, either directly or through the effects of toxins. Once the host is dead, they sit around and wait for a carnivore to eat their hosts and digest them. If lucky enough to get inside the stomach of the new host, the digestive juices break down the larvae's coverings. This cues the larvae to dig a hole through the intestines and burrow into flesh, becoming adults and giving birth to a new generation of parasites.

The tapeworm and *Trichinella* that our hunter-gatherer ancestors had to deal with, however, were a cakewalk compared to the infectious diseases in store for our agricultural ancestors. To begin with, these were parasites that had previously lurked in the surrounding environment but now multiplied due to poor sanitation, close quarters, large populations of humans and domesticated animals, and human-made habitats. Malaria, filariae, yellow fever, and dengue fever were carried by mosquitoes that hung around like tiny thugs in shaded areas around houses and animal shelters, laying their eggs in convenient urban puddles and reservoirs. The diseases they bore came from wild animals, particularly primates. The protozoan that causes malaria was originally a parasite of macaque monkeys and chimpanzees. The yellow fever flavivirus was a longtime resident in West African monkeys, while the closely related dengue flavivirus spread from monkeys to people living in Asia, and perhaps competed with yellow fever.[42] From the viewpoint of parasites, a dead host is usually not a congenial one, so malaria and dengue fever, both long-established diseases, tended to result in low death rates, albeit with painful and dangerous episodes of fever and liver damage. Humans evolved genetic adaptations to deal with the dangers of malaria.[43] Yellow fever, on the other hand, moved relatively recently out of West Africa with the slave trade that began in the seventeenth century and delivered high mortality rates of up to 80 percent in newly exposed populations. Dengue fever has existed for an intermediate period and has a maximum mortality rate of 50 percent.[44]

A few years ago, after returning to Vietnam from a motorcycle trip through Thailand and Laos, I came down with fever, chills, and diarrhea—dengue fever—in Hanoi. I checked myself into the Vietnam-Cuba Friendship Hospital, then spent a few miserable nights in the crowded malaria-dengue ward lying on a metal bed with another man's feet in my face. The open-pit toilet in the adjoining room reeked, but I was so weak that I couldn't even hobble there without fainting and had to make use of a bedpan—humiliating when you're crammed in a ward with a dozen patients and their numerous relations. To add insult to injury, a would-be-thief snuck in among the billowing mosquito nets during the middle of the night, aiming for a wallet under a man's pillow, and I had to shoo the intruder out of the ward. It was a miserable, debilitating experience, but one that woke me up to the real dangers of parasitic diseases and the challenging conditions that most people in the world have had to contend with, including our ancestors.

Aside from mosquito-borne diseases, other parasites and parasitic diseases that humans picked up during the transition from nomadism to often filthy and crowded sedentism include smallpox, originally harbored among monkeys; plague, from rats; typhus, from rats, lice, and fleas; schistosome flukes, borne by snails in slow-moving waters; sleeping sickness, transmitted by tsetse flies; leishmaniasis, or river blindness, from sandflies; Chagas disease, carried by ticks; and rickettsioses, from ticks, mites, and chiggers. Other pathogens have unknown ancestry: bacteria like the ones that cause leprosy, rubella, whooping cough, diphtheria, chicken pox, syphilis, cholera, poliomyelitis, meningococcal meningitis, and hepatitis, as well as shigella, *E. coli*, and streptococcal and staphylococcal bacteria; viruses like the ones responsible for viral encephalitis; and fungi like *Candida albicans*.

Then there were the parasites that we caught from our newly domesticated animal companions. During my undergrad years when I went backpacking through South America I disdained tourist restaurants and preferred street food. One night in Ecuador, I bought barbecued pork from a street-side stand. The slab of meat, hung from the roof, was lit with a single naked bulb, and a horde of flies buzzed through the smoke. I watched the serving lady dip my fork and dish into a bucket of gray water mottled with scum before the utensils were handed to me. For some unknown reason, I still ate the pork.

Two days later, I set off into the Ecuadorian rainforest, following a path that, judging from a crude map, I guessed would lead to somewhere interesting. I had strong legs, a sturdy backpack, a tent and stove, and several packets of dried noodles. After about an hour of trudging over roots and slipping farther and farther away from civilization, I was beset with a strong thirst and a headache pounded at my temples. I dropped to my knees. Damn, time to backtrack. I made it to a hotel at the edge of the jungle, where I boiled a bowl of noodle soup in the hotel kitchen, then fainted. When I awoke, I found that I had smacked my head into the bowl of soup, sending its cheap contents all over the table and floor. Hotel staff helped me get back to my room on the second floor, where I swooned through a high fever that afternoon.

The next day, I took a bus and then a cab to the closest town and found a hotel. I lay in bed, racked with fever and diarrhea. The following morning I took a cab to the state hospital. I blurted to reception, "I have malaria!" and was rushed to a bed and administered an emergency dose of mefloquine, the antimalarial medication that I had been taking on a regular basis. The preventative dosage was once a week, but the doctors now gave me three or four pills in quick succession. The doctor leaned over me and asked in clear English, "Do you know anyone in this town?"

"Someone named Maria. She owns a restaurant," I responded weakly.

The medicine caused me to be extremely nauseated. I couldn't eat any food for a few days and wobbled unsteadily to the bathroom. It also induced other effects. I became absolutely convinced that the male nurses were going to kill me. The only question was whether it would be accomplished through dropping a stone on me, knifing me, or slipping poison into my medication. At the same time, I was absolutely convinced that one of the student nurses had fallen in love with me (they were all teenagers) and thought I could hear her voice mingled with her friends', chattering about the Canadian down the hallway. When I closed my eyes, I saw dismembered heads marching against a black backdrop, climbing into a tree.

Somehow the hospital staff managed to track down Maria, a restaurant owner whom I had briefly met and chatted with on my way into the jungle. Maria used to work for a citrus company in California and had retired. She spent half her year in California and the other half running her hotel/restaurant/farm on the edge of the Ecuadorian rainforest. "I am the happiest person in the world!" she had declared to me, and I believed it. Maria contributed to the local economy by providing work for the local youth. Now she came into my hospital room bearing a fragrant baked spring chicken, slaughtered that morning for my meal, along with coleslaw and fries—classic American comfort food. I was extremely touched by her generosity and thoughtfulness.

"Don't you worry, they'll take very good care of you here," she remarked.

I hadn't been able to touch any of the hospital food prior to Maria's visit—mostly rice and beans—but something in Maria's magical baked chicken roused my appetite. I devoured her meal. She sat back, watching me with a smile. The next day, I was able to get up and walk around the hospital. A man in my ward, amazed at my rapid recovery, gestured at the thick book that I was leafing through—the *Lonely Planet* guide to South America.

"Is it the Bible?" he asked.

I didn't have the heart to tell him the truth; besides, it was my bible, in a way. I told him I was interested in religion, so he came to my

bedside and prayed fervently, thanking God on behalf of the idiot athe-
ist. When I checked out of the hospital, I tried to pay for my stay, but
it was a state hospital where all were admitted for free, even foreign-
ers. I shoved some bills into a doctor's hands and asked him to do some
good with it. I dropped by Maria's restaurant. She beamed at me but
also would not take any money, stating simply, "Do something good
for someone else in return."

When I got back to Canada and told my doctor about my pre-
sumed bout of malaria, he said that it was much more likely to have
been food poisoning caused by typhoid; malaria would have produced
more severe symptoms and taken far more time to recover from.
Typhoid is caused by *Salmonella typhi* and likely arose from our post–
Agricultural Revolution association with domestic animal poop and
dirty water, which allowed the bacterium to eventually transition from
living in domestic animals to our guts. *Salmonella typhi* are crafty enough
to evade detection by the immune system while nesting within our
cells. When they burst out, the immune system goes berserk, prompting
the bacterial horde to pump out endotoxins. This probably triggered
the misery that I had experienced in Ecuador: severe headache, weakness, aches and pains, fever. I was fortunate to escape from the clutches
of typhoid relatively lightly; 10 percent of people afflicted with typhoid
fever die.[45]

Infectious diseases like typhoid and influenza, and parasites like
hookworms and pinworms, arose from comingling with unspecified
domesticated animals, but others have been traced to specific ancestry. Dogs are perhaps our longest-running animal associates—we go
back at least ten thousand years together—providing us with watch-
ful sentinels, hunting assistance, and possibly dog stews and steaks (as
still practiced in many parts of the world, particularly Asia), but also
rabies, whipworms, and our first introduction to the deadly scourge
of measles. Cows gave us their milk, their meat, and their tuberculosis, as well as the much-feared anthrax. Goats furnished us with milk
and brucellosis, also known as Mediterranean Disease, characterized
by fevers, chills, weakness, headaches, depression, and weight loss;

Napoleon Bonaparte suffered from symptoms consistent with brucellosis, and the diagnosis was confirmed in tissue samples taken from his body. Poultry gave us drumsticks and the mumps. Cats dispatched bothersome rodent pests but also infected us with the protozoan disease toxoplasmosis. Perhaps one in two Americans carries the protozoan parasite *Toxoplasma gondii*, which is usually harmless but can inflame lymph glands in the neck and cause low fever and fatigue, or, in people with compromised immune systems, severe damage to heart, muscle, and brain tissue. Horses were difficult to tame, but after humans domesticated them around three thousand years ago, they returned the favor by bequeathing us the common cold. Pigs gave us heavenly barbecue, along with the giant roundworm *Ascaris lumbricoides*.[46]

Roughly one out of every four people in the world is infected with *Ascaris lumbricoides*. This parasite begins its bizarre life cycle in the small intestine of an infected person, where the hardworking mother worm, up to a foot long, pumps out 200,000 to 240,000 eggs per day. Her progeny exit the body along with feces and, if they reach soil, take about one month to develop to the infectious stage. If an egg is unwittingly swallowed (children at two institutions in Jamaica were found to have ingested on average nine to twenty giant roundworm eggs a year), it hatches in the small intestine of the new host.

Now, if the giant roundworm larvae were content to just grow up, mate, and produce eggs in the small intestine, then infections with these parasites would be more benign than they actually are. Instead, like college kids with a yen to backpack through Europe or Asia, the larvae head out into the great unknown expanses of the human body for a sightseeing tour. First, they burrow through the lining of the small intestine and enter the circulatory or lymphatic system. The young parasites journey to the lungs, spend around two weeks fattening up in the lung capillaries, break into the alveoli air sacs, wander through the lower respiratory tract, shimmy up the larynx, and are then coughed up and swallowed by the host. This puts them right smack back where they started, in the small intestine. So why go through so much trouble? After all, from a biological viewpoint, all that effort of traveling around the human body is wasted, like calling a moving company and having them transport your belongings around the country for a few months before moving back into the same home.[47]

The likeliest explanation is that the small intestine is a pretty harsh place to live, even for a parasitic worm. You're immersed 24/7 in a scalding bath of gastric acid, bile, and digestive enzymes; the intestine walls are always trying to push you farther downstream; the mucous membranes that you're trying to hang on to have a tendency to slough off like mountain avalanches; a barrage of bulky human food pummels you; and oxygen levels are chronically low. By comparison, life in the tissues is a Mediterranean resort. The immune system can't hammer an invader ensconced in lung tissue the way it can blitz intestinal pathogens, because lung tissue is especially sensitive to inflammatory attacks. Migrations of parasitic worms out of the small intestines is a change of scenery that provides a chance to grow larger and hence more reproductively robust.[48]

Most giant roundworm infections do not cause great harm to their human hosts—nor should they, because if they were deadly, we wouldn't be here and neither would they. However, during their epic peregrinations about the body, they may get lost, and that's when the real damage can occur. Giant roundworms have been found in the sinuses, pancreas, bile duct, gallbladder, liver, lower intestines, and appendix; they have caused cardiac arrest; they have exited the body through the ear and vagina, and from the bladder with urination. They can even infect unborn babies; a foot-long male roundworm was once recovered from a newborn child. Giant roundworms can be extracted or expelled with drug treatments, but reinfection is common. Infants, due to their fondness for eating soil, are easily contaminated. Giant roundworm eggs are virtually indestructible, resisting acid, alkaline, dehydration, and toxic salts.[49]

It's worth contemplating the life of a roundworm, not only because they are a health hazard for billions of people around the world, but because their life cycle helps to explain how the hygiene hypothesis works. If you have the misfortune to acquire a highly specialized parasite, the last thing your body should do is to mount an extremely aggressive defense. The giant roundworm has coevolved with humans for thousands of years, so it is highly resistant to purging, and an inflammatory barrage strong enough to quickly knock out the roundworm would end up destroying innocent surrounding tissue. Instead, the immune system response is toned down to a level more appropriate for a long siege with a wily foe, and as a side effect, other foreign substances, such as grass pollen and cockroach dust, also get treated with kid gloves instead of brass knuckles.[50]

In other words, long ago in our evolutionary history we struck a kind of bargain—a truce among thieves—with parasites: In return for their not killing us (or not too quickly, at any rate), we grudgingly agreed to host them in our guts and feed them with our blood. Paradoxically, our immune systems became dependent upon early bouts of infectious diseases for proper calibration. Without the crucial intervention of infections, our immune systems remain immature and oversensitive. Truce or no truce, no one likes dealing with thieves, especially the kind that reside in our internal plumbing. A great sweep of hygienic innovations, inspired by the work of Pasteur, Koch, and other scientists—notably the widespread use of antibiotics—virtually eliminated parasitic horrors like giant roundworm from everyday life in industrialized countries, but in its place, spotless households are besieged with allergic diseases, and our former parasites are getting the last laugh.

The hygiene hypothesis, it should be noted, is far from perfect. Why, for instance, do some studies show that infections from a particular parasite alleviate allergic disease symptoms or markers, while other studies on the same parasite find the opposite result? Why do some parasites produce fewer inflammatory symptoms than others? Why are asthma rates decreasing in Western countries?[51] Not because environments are becoming dirtier or because families are getting bigger. As for food allergies, no one knows yet if the hygiene hypothesis applies to them; so far the strongest results in allergic diseases have been with asthma and hay fever.[52]

Despite the many gaps in the hygiene hypothesis, researchers have found that therapeutic infection with parasites can produce impressive improvements in patients with chronic autoimmune diseases. Crohn's disease is an inflammatory disease of the intestines, leading to fever, abdominal pain, diarrhea, weight loss, vomiting, rectal bleeding, arthritis, and painful nodules on the shins. Researchers at the University of Iowa in Iowa City asked participants with Crohn's disease to down a drink with 2,500 pig whipworm eggs floating in suspension, every three weeks, for a total of twenty-four weeks. Pig whipworms were chosen because, unlike their human whipworm relations, they do not infect humans naturally, they hatch in the gut but politely refrain from venturing beyond their environs, and they are expelled after a brief period of colonization. Within three months, almost all of the Crohn's disease sufferers who drank pig whipworm eggs went into remission. Because the participants knew what they were drinking, the researchers could not rule out a placebo effect, but even then the percentages were promising. Dr. Joel Weinstock, one of the authors of this study, points out that Jews may have higher rates of Crohn's disease because they refrain from eating pork, have cleansing rituals, and tend to live in cities far from animals and animal waste, which may displace them from the protection offered by parasitic worm infections.[53]

Ulcerative colitis, a disease that shares many of the features of Crohn's disease (both are grouped under the heading of inflammatory bowel disease), was also the subject of an experimental study by researchers at the University of Iowa and again produced positive clinical improvement in symptoms. Buoyed by the promising results of these small trials, larger-scale tests on the safety of parasitic worm therapy are now being carried out in Europe and the United States. Other researchers are investigating the applicability of using parasites to battle multiple sclerosis and type 1 diabetes. One day, Mom, Dad, or Doc may pop a couple of thousand FDA-approved whipworm or hookworm eggs into your apple juice ("a worm a day keeps the wheezing away . . .") to go with your breakfast. Until then, however, some scientists worry that the message about the hygiene hypothesis is going in the wrong direction and risks undoing decades of public health measures to keep serious infectious diseases at bay.

Vaccines are of great interest in this regard. The logic of the hygiene hypothesis suggests that vaccinations should increase the risk of allergic diseases by removing childhood infections. Although there is preliminary evidence for this possibility—vaccinations against a species of bacteria responsible for respiratory tract infections may increase the chance of getting asthma, whereas chicken pox infections in early childhood may reduce the likelihood of eczema and asthma—most doctors strongly support vaccinations, arguing that the dangers of measles, mumps, chicken pox, and other classic childhood diseases greatly outweigh the risk and inconveniences of allergic diseases.[54] Some parents may fret about the consequences of mercury and aluminum in vaccines in promoting diseases such as autism, but the 1998 study that originally sparked questioning about the link between MMR (measles, mumps, and rubella) vaccine and autism has been thoroughly discredited for manipulation of data, and the principal author of the study, the British doctor Andrew Wakefield, was stripped of his license to practice in the U.K. in 2010. Nonetheless, many people remain confused about the safety and efficacy of vaccines, which has led to an alarming jump in the prevalence of once-uncommon contagious and potentially deadly infectious diseases like measles and whooping cough among children in industrialized countries.

Ironically, while many scientists worry about the sharp rise in preventable infectious diseases, other individuals worry that bureaucratic inertia is bogging down the release of parasitic medications that could rapidly alleviate their serious allergic and autoimmune diseases. For a few thousand dollars, a person can cut the red tape and buy pig whipworm or hookworm eggs online today from shrewd enterprising companies. Down the road, when FDA-approved parasitic medications finally hit the market, they will provide a measure of safe relief for allergy sufferers.[55]

These kinds of regulatory innovations will take time to implement. In the meantime, for parents concerned about the possibility of allergies, giving children more exposure to sunlight, lessening dependence on the use of antibiotics and antibacterial hand soaps, and achieving a better balance of omega-3/6 fatty acids through increased consumption of animal fats and/or decreased consumption of omega-6-heavy vegetable oils (such as corn oil) are practical measures that plausibly decrease the risk of allergic disease.

Meanwhile, to prevent common infectious diseases like measles from regaining a foothold in our societies, parents should continue to follow the advice of doctors and vaccinate their children. The case of vaccinations is a good example of the limits of applying evolutionary theory to everyday health: Although our bodies were adapted for exposure to nasty bacteria, viruses, and other parasites, it simply doesn't make sense to reinvite that nasty crowd back into our living rooms. It is more sensible to give scientists time to find harmless parasites to calm our immune systems down, and to let our children hang around farm animals in the meanwhile.

# THE CALORIE CONUNDRUM

[In a 1987 study by Drewnowski and Yee] 90% of American boys were dissatisfied with their weights, the same number as among girls, but whereas all the girls wanted to weigh less, half the boys wanted to weigh more.

—CLAIRE M. CASSIDY, *"The Good Body:*
*When Big Is Better"*

Ravens dot the wintry sunset over Sapporo. I scurry past precipitous snowbanks and walk-skate across a treacherous icy street. A staircase leads down into a shopping mall under the central train station. Shop windows gleam with fastidious replicas of Japanese delicacies: lacquered bowls bearing buckwheat or rice noodle soup and slices of fatty pork, rice topped with bright yellow omelets and rivulets of mayonnaise or ketchup, pork bones moored in curry stews. My stomach growls at the odors wafting out of the shop entrances, but because the first installment of my university researcher's salary will not be paid out until the end of the month, I keep walking, looking for something cheap and filling.

I start to feel woozy from pacing up and down the corridors on an empty stomach. From the depths of my despair, I spy a lavish poster: In shiny pictures—for such is the extent of my command of the Japanese language—it promises a three-course meal, including miso soup, a salad, and a bowl of rice topped with salmon roe, for only 700 yen, or

$8.50! I nearly weep in relief as I march into the restaurant, unperturbed by the upscale office workers gathered inside, the posh wood furnishings, the waitress smartly dressed in a traditional outfit. An advertised bargain is iron-clad, a penny saved with a hot meal to boot. Isn't life sweet? After I squeeze onto a free chair at the bar, I am handed a leather-backed menu. Hmm. Most budget eateries that I frequent offer menus consisting of plasticized home-office printouts, but no matter, a special is a special. I flip through the pages but cannot find the meal combo advertised on the window. I keep flipping through the pages. It is hot inside the restaurant; I am still wearing my bulky, frayed, fifteen-year-old industrial-green winter parka. I search in vain through the pricey offerings. I gesture at the window and explain in my best broken English, "Food–special–window–outside?"

The waitress shoots back without a hint of a smile, "Lunchtime only."

Oh dear. Far too late to make a discreet exit. Two women have ceased their chatter to survey the foreigner in an awkward predicament. Sweat starts to pour down my back and forehead, and my armpits begin to release alarm pheromones. What to do? Trying to mask my agitation, I flip through the pages again, front to back, back to front. I point to the cheapest offering I can find.

An Osaka salaryman next to me strikes up a conversation. Presently, the waitress places in front of me a fist-sized serving of rice, topped with a smattering of salmon roe. The salaryman and I both stare in astonishment at my dinner. He, evidently, is traveling on a business expense account, for he has ordered the same rice, plus a bowl of soup, a side of marinated fish, a salad, a frothy omelet, and a tall bottle of Sapporo beer. I fantasize what my restaurant meals would have been like if I had sold my soul to the corporate world instead of miserly academia. The salaryman asks with a note of incredulity, "That's all?"

"I'm not hungry," I reply, trying to sound convincing.

As I discovered after moments of humiliation like this, food in Japan is expensive, due to a confluence of high transportation and labor costs, limited growing season, shortage of arable land, and import barriers. At the other end of the cost spectrum, Americans enjoy access to the cheapest food in the world relative to income.[1] True, bargains in Sapporo can be ferreted out here and there, like generously proportioned *okonomiyaki* pizzas made of cabbage, dough, and squid, with a squirt of mayonnaise, served up by a trim sixty-year-old chef just beyond the university gates, or brisk ramen noodle shops and rice bowl joints that take orders through vending machines that dispense meal tickets. In Los Angeles, on the other hand, when my college friends and I were hungry, we could order a heaping, steaming, savory plate of noodles, pork, and veggies, plus a beer, in bustling Thai Town; stuff our faces with pancakes, eggs, and sausage at Denny's; grandly feast on Mexican-style rice, black beans, lettuce, shredded pork, and salsa at Chipotle's; or fill our stomachs on a never-ending parade of salad, soup, pasta, pizza, sourdough bread, baked potatoes, bland apples, and ice cream at Souplantation, all for an astonishing ten bucks or less. No wonder the average Japanese man eats around three hundred fewer calories per day than his American counterpart, less than even the Chinese. The Japanese also have longer life spans than Americans and Chinese. Could the smaller food portions of Japanese have something to do with this?[2] In the never-ending discussion of diet and health, the debate often comes back to calories. Is the amount we eat killing us? The answer is surprising and may have a profound impact on the way we choose to eat and live.

In a succession of experiments carried out since the 1930s, it has been observed that reducing the amount of food that animals eat causes some species to live longer. This "calorie restriction" effect has been found in diverse species. In 2009, the prestigious peer-reviewed journal *Science* published an article that seemed to seal the case for human participation in calorie restriction: In a two-decade-long Wisconsin study, monkeys that had been allowed to eat as much as they wanted died from diabetes, heart disease, and cancer at a faster rate than monkeys that had been allowed to eat just 70 percent of the calories permitted to the first group. Is this a no-brainer? Overeating kills, so is it time to get serious about cutting the calories? Many people have already reached this conclusion and voluntarily maintain a diet of reduced portions. However, there are a few wrinkles in this line of thinking. Although many scientists endorse the principle of calorie restriction, others are skeptical because the evidence on calorie restriction with humans has been minimal.[3]

Meanwhile, Shinichi Nakagawa and his colleagues at the University of Otago in New Zealand perused the results of more than a hundred calorie restriction experiments and noticed four surprising themes. The first is that the longevity benefit from reducing calorie consumption has been demonstrated mainly in animals that were bred for laboratory conditions: rats, mice, fruit flies, and yeast. Many wild animals have been tested, including fish, grasshoppers, and moths, but these don't show the same dramatic improvement in life span when their food portions are reduced. No one knows why this is the case, but lab animals live in a peculiar world where food is never scarce, and thus some important pathway in their physiology may have been altered or lost through generations of controlled breeding. A lab animal's appetite is uncoupled from the demands of life in a natural setting. By contrast, wild animals may be like Swiss timepieces, exquisitely honed by evolution to eat the right amount of food. It could be argued that humans in industrialized regions have also been exposed for many generations to conditions where food is in abundant supply, and therefore the calorie restriction effect might still apply to us—in other words, our genes may have more parallels to the genes of lab-bred rats and mice than to those of wild animals.[4]

The second wrinkle of the calorie restriction effect is that the life-lengthening result of calorie restriction is overshadowed by an even stronger impact from protein reduction. In other words, if you're looking to pump more days into your life, cutting back on calories while increasing your protein intake may end up accomplishing nothing, but keeping your calorie intake constant while reducing meat and other protein sources may be the clincher. Reducing protein intake causes a decline in IGF-1 circulating in the human bloodstream, which may turn out to be a good thing, because IGF-1 has been linked to increased risk of prostate and premenopausal breast cancer.[5]

A third recurrent theme is that a reduction in calories and protein brings benefits only up to a certain point; after that, if calories and protein continue to be eliminated, the organism will begin to suffer adverse health. There exists a sweet spot, an optimal intake level of calories and protein where life span is maximized. In studies, cutting calories to half of the organism's preferred intake and slashing protein by two-thirds yields longest life. Remember, these are estimates carried out across all species, including the laboratory animals that exhibit the strongest results from calorie reduction. For humans, the optimal levels of calories and protein may turn out to be different.[6]

The fourth issue to consider is that females tend to reap more benefit from calorie restriction than males. This amply documented sex difference has a worrisome implication: Cutting back on calories probably causes sexual desire to evaporate. Proponents of calorie restriction (they seem to be mostly men) are understandably less than eager to draw attention to this effect. The best argument thus far about why calorie restriction increases longevity is that extended hunger causes animals' bodies to switch priority from reproducing to prolonging life. It's analogous to a bear's instinct to hibernate through a long winter rather than waste energy by lumbering around a snowy forest searching for nonexistent food and mates. Many scientists believe that food deprivation similarly triggers a diversion of energy away from hopeless activities such as trying to conceive when a mother has barely enough food reserves to keep herself alive, instead channeling those scarce calories into repairing the body and conserving energy for a better opportunity down the road, when food finally comes hopping down the path or sprouts from a branch.[7]

For females, holding off on babies yields a jackpot of savings in energy. For males, on the other hand, having sex now or later does not entail a dramatic shift in physiological effort. The bodies of males may therefore be geared to smaller increases in longevity when food intake is reduced.

As you can see, the edifice of this theory is erected, so to speak, on the axiom that calorie restriction shifts priorities from reproducing now to reproducing later. Besides draining sexual urge, half-starving yourself will transform you into an ill-tempered ogre. Not surprisingly, taking food away from animals makes them aggressive. I remember an occasion when my family was tardy in feeding the house cat. When I finally brought out its plate, rather than being grateful, our normally placid cat leapt at me and scratched my leg. If people who are well fed view society on benevolent terms, as they are forced to forgo greater quantities of food, their circle of sympathy will steadily shrink, from society to friends to family to close family, and finally to the self alone.[8]

To rub barbecue salt into an open wound, the latest round of news for calorie restriction boosters is a tad gloomy. A second set of monkeys is currently being tracked by the National Institute on Aging (NIA) in the United States, and the results so far, released in the fall of 2012, indicate that shaving calories didn't help hungry monkeys gain extra years compared to well-fed monkeys. However, this most recent NIA experiment compared fit versus scrawny monkeys, while the previous Wisconsin study compared overweight versus scrawny monkeys. There may be only a slight difference in longevity between being fit and being scrawny (both are relatively healthy states), but a considerable difference between being fit and being overweight.[9]

Even if you'd rather eat ice cream while scientists quibble and the members of the Calorie Restriction Society practice dietary austerity, you may want to consider putting your dog on a diet. In a study of Labrador retrievers, half were allowed to eat until they were sufficiently (but not over-) fed, while the other half were allotted three-quarters this amount of food. By the time all the well-fed Labs had passed away (the last one survived to thirteen years, a ripe old age in dog time), nearly 40 percent of the underfed dogs were still alive (and irritably waiting, one presumes, to be fed).[10]

Although critics argue that the longevity benefits from calorie restriction are underwhelming, almost all scientists agree that the physiological effects are generally positive: These include reduced incidence of the most common human chronic diseases (diabetes, cardiovascular disease, and cancer), slowdown in cognitive decline, and lower levels of cholesterol, triglycerides, glucose, and insulin. The chief drawbacks are that a calorie-restricted animal stops growing, becomes less fertile, and is more vulnerable to cold and some infectious diseases.

Taken to an extreme, calorie restriction imposes psychological and physiological side effects that few people would be willing to tolerate, but even a modest 10 percent reduction in calories—from buffet-style consumption to eating just enough to maintain constant body weight—is a supremely cost-effective ticket to a bonanza of health benefits.[11]

Although today the chief concern of many people in industrialized societies is reducing calorie consumption, for other places in the world the great struggle is avoiding starvation, and this was even more true in the past. This observation may seem banal, but an examination of the history of calorie consumption turns out to be fascinating. By examining this history, we can gain a better idea of why many people today struggle with health problems linked to calorie consumption, particularly obesity.

A few thousand years ago, if gamblers had wagered on which society would first beat the scourge of famine and rack up the caloric intake, the safe choice would have been the Chinese, based on their vast agricultural knowledge. The Chinese knew how to treat poor soil with organic wastes, ashes, manure, human waste, and river silt.[12] By AD 0, they replaced slash-and-burn agriculture with complex crop rotations.[13] They knew how to mix crops using plants like broad beans and ferns, and by the sixteenth century, they knew about the application of potash (minerals that contain potassium) and oil cake (the residue from seeds pressed for oil). Their authorities advocated plowing in the burned stubble from harvests. Through meticulous experimentation and refinement of farming practices, China was able to support a population of more than 100 million people in AD 1124; in comparison, the population of England, at under a million, was around the size of a large Chinese city.[14]

At first glance, the path of agriculture in the West seemed similar to China's, albeit slower. Through trial and error, and observation, the Romans learned to employ chalk, dung, and ash and to intercrop lupin (a kind of legume), beans, vetches, and clover. After the collapse of the Roman Empire, crops were changed from two-field rotation to three-field, then four-field rotations of corn, clovers, grasses, and fallow cropping. Inland oceans and rivers provided convenient highways for commerce, and the riches gained from trade propped up a social class that was interested in further profit and the technical means of increasing that profit. The decimation of the European population from the Black Death in the fourteenth century—likely introduced by rats from China, ironically—broke the static pattern of manorial serfdom and freed the gentry to exploit their lands for profit. The disparity in wealth also guaranteed that while some people worked continuously and lived in poverty, others had the time and means to engage in scientific inquiry.

Despite the great effectiveness of Chinese agricultural techniques, Chinese knowledge had accumulated from trial and error and sharing of knowledge over the generations. There was no sustained effort in China—or anywhere else in the world, outside of Europe—to understand why these techniques worked, to discover what ashes, manure, broad beans, ferns, potash, and oil cake all had in common.[15] Scholars were revered in China, but the scholarship concerned social relationships and was viewed as a means to gain access to prestigious mandarin employment, while commercial activities were disdained. Moreover, contact with other civilizations was relatively limited, due to barriers of mountains and long distances. Virtue, in the eyes of the Chinese, was found in honest rulers, stable societies, filial piety, hard work, and thrift. These were also the pillars of medieval Europe, but many of these lessons were forgotten as Europe turned toward capitalism and science.[16]

The key stumbling block to increasing agricultural yield was developing a proper scientific theory of the elements and of nitrogen in particular. One prominent step toward creating such a theory came from the Flemish scientist Jan Baptist van Helmont, who grew a willow tree from 5 pounds to 169 pounds in five years' time, with nothing but water added to the plant and soil. The soil decreased in weight by just two ounces, which he believed indicated that the tree had somehow transmuted water into tree material.[17] An excellent experiment; the conclusion was wrong, but the method was precise, and other scientists would be able to make progress. A critical step was to understand what the substance was in air that helped plants grow and what this substance had in common with materials like beans and dung. Like a net around a fish, the scientific theories connecting plant-promoting substances were drawn together in a frenzy of collaboration and competition among European scientists. In 1772, the Scottish chemist Daniel Rutherford succeeded in isolating nitrogen gas. The English scientist Henry Cavendish applied an electric spark to a mixture of oxygen and nitrogen gases, which produced nitric acid; this was combined with sodium hydroxide to create a solution of potassium nitrate. The extraction of nitrogen from the atmosphere was one of the most important intellectual breakthroughs in the history of humankind; without this discovery, the population of the world today would have remained close to the 1800 level, around a billion people.[18]

Extraction of nitrogen in a laboratory was one thing; ramping up the conversion of nitrogen into various forms of ammonia as artificial fertilizer to help feed the hungry masses was another. At the opening of the nineteenth century, there was no practical way of achieving large-scale nitrogen fixation with the crude technology that was available. By 1913, two German chemists, Fritz Haber and Carl Bosch, and the industrial juggernaut BASF had surmounted the technical challenges. For Germany, the timing was auspicious, because the outbreak of World War I in 1914 led to the severing of the British-controlled guano-nitrate supply from Chile, which had been essential for the manufacture of explosives.

In complete contrast to the gargantuan supply of energy (from hydropower, coal, and, these days, "natural" or methane gas) and physical infrastructure required in industrial nitrogen fixation, there is another way to extract nitrogen from the atmosphere and convert it into usable form. Legumes like peas and beans maintain a form of bacteria in their root nodules. In soil, the rhizobia bacteria are free-living souls, like a crowd of laid-back hippies. When they get a chance to migrate into root nodules, however, they transform into bacteroid structures, roll up their little sleeves, and get to work. The legume plant protects its oxygen-sensitive bacteroid workers by removing oxygen and feeding the rhizobia bacteria meals of glucose. In exchange, the rhizobia release phosphate and energy; the energy is used by the bacteria to split the bond between the two atoms of nitrogen gas, freeing the nitrogen to combine with hydrogen and become available for the plant to use in the form of ammonia.[19]

What is most remarkable about the rhizobia is the extraordinarily scant energy required by the bacteria to split the powerful bond binding nitrogen molecules. Industrial furnaces used to fix nitrogen require temperatures well beyond the range produced in ordinary fires, and thus electric furnaces with special heat-resistant protection are essential to industrial nitrogen fixation. Modern science has not yet figured out how the rhizobia work their nitrogen-splitting magic with such energy thriftiness. Some twelve thousand species of *Leguminosae* perform nitrogen fixation, though fewer than fifty of these species are employed in agriculture. It is extremely humbling to think that a tiny bacterium can accomplish with ease what humans require massive inputs of energy and complex furnaces to do.

Nevertheless, since coal and natural gas deposits have been relatively plentiful, the process of fixing nitrogen using the Haber-Bosch method is feasible, for rich nations at any rate. Before 1840, no inorganic nitrogen had ever been applied to crops. A hundred years later, more than 3 million tons of nitrogen were being applied to farmlands every year, with more than three-quarters of this amount coming from industrial production. By 1988, production of nitrogen had increased nearly thirty-fold. This massive increase in synthetic nitrogen and the food it yielded meant more people coming into this world, surviving to older ages, and getting bigger in size. The use of nitrogen fertilizer around the world is greatly skewed, however; very little fertilizer, for example, is applied in Africa, while huge amounts are used in the developed countries to support the raising of livestock, which are notoriously inefficient at converting nitrogen into human food. This is one reason why daily calorie intake varies so widely around the world.

If the Europeans take credit for the discovery of nitrogen and the subsequent explosion in food availability, they also deserve credit (or blame) for another technological innovation that had a profound impact on calorie consumption. For millennia, humans had been trying to solve the inconvenience of walking. Wheeled sleds were employed in Sumeria in 3500 BC. The Egyptians had horse-drawn chariots by 1600 BC.[20] The ancient Romans constructed an extensive network of roads for their chariots. By AD 1650, London suffered from traffic jams caused by horse carriages, despite the considerable discomfort of riding in these contraptions. Passengers paid by the mile; inside seats cost twice as much as outside seats. Horses were changed frequently on long trips.

The first self-propelled automobile was invented in the eighteenth century, and ever since people have been obsessed with cars. In the twentieth century the United States became a global leader in churning out cheap cars for the masses. City planners all over America adjusted for the influx of cars by building sprawling suburbs that catered to, and required, a car. Not only in the United States but all across the world, people were changing their habits to accommodate the vehicle that everyone wanted to own.

At this point, a reader might conclude that the root of modern food-related ailments like obesity and diabetes lies in people eating a lot more food, due to the miracle of nitrogen fixation, and doing a lot less physical activity, due to the miracle of combustion engines and private vehicles. However, it turns out that neither of these common beliefs is supported by the evidence.

First, the food intake myth. The daily energy consumed through food in contemporary industrialized nations runs from about 2,300 kcal (kilocalories) among Japanese men and 1,800 kcal among Japanese women to 2,600 kcal among American men and 1,900 kcal among American women.[21] What is surprising is that the average daily caloric intake of these overweight industrialized societies is about the same as among hunter-gatherer groups, with some hunter-gatherer groups below and others above the calories consumed of industrialized nations.[22] Although hunter-gatherers ate about as much as we do today, they faced much greater variability in their food supply.

In northern Australia, among the Anbarra, the daily energy intake dropped to 1,600 kcal during the rainy season and peaked at 2,500 kcal during the dry season. The calorie consumption of the Hiwi in the rainforests of Venezuela bounced between 1,400 and 2,800 kilocalories, depending on the season (plant foods were most plentiful at the end of the wet season). Thus, if any major pattern emerges in terms of caloric intake, it is that our hunter-gatherer ancestors lived on a dramatically varying diet, which swung between feast and famine according to the season and other hazards of fortune.

Another surprising finding concerns physical activity. Although it is commonly believed that people in hunter-gatherer societies expended much more energy than people in industrialized societies today, the evidence so far does not support this assumption. One common measure of physical activity level (PAL) expresses the total energy used in one day as a multiple of a person's metabolic rate. For example, a PAL of 1 means that a person uses only his/her metabolic energy, i.e., the energy expended by breathing, thinking, digesting, etc. A PAL of 2 means that a person uses twice as much energy as his or her base metabolic rate. PAL allows us to adjust for the fact that people have varying levels of metabolism; a person who has a high metabolic rate can burn up a lot of energy by just sitting in one place compared to a person with low metabolism, so a good measure of physical activity needs to compensate for differences in metabolism. To determine the amount of energy used in a day, the best measure involves giving a person a drink of water that has been "tagged" with isotopes of hydrogen and oxygen. Measurement of these two tags in samples of saliva, urine, or blood allows measurement of exhaled carbon dioxide and hence the degree of respiration from metabolic processes.

Using tagged water, the average PAL among foragers was found to be 1.78 for men and 1.72 for women. Among industrialized contemporary societies with a high human development index (which measures income, literacy, and so on), the PAL of men was 1.79 for men and 1.71 for women.[23] In other words, the energy expenditure of overweight contemporary industrialized societies is roughly the same as that of lean hunter-gatherer societies once metabolism is taken into account; or to put it another way, the cause of obesity is unlikely to be lack of exercise, because people in industrialized societies today use about the same amount of energy as people in hunter-gatherer societies.[24]

This finding has important implications for understanding obesity. All of us living in industrialized societies are aware of the stigma associated with obesity, and perhaps the longer-term health consequences of diabetes, high blood pressure, gout, and cancers associated with being overweight. Since food intake and energy expenditure levels today are roughly the same as during ancestral times (using the lifestyles of modern hunter-gatherers as a reasonable model for our ancestors' lifestyles), why are obesity and diabetes so prevalent among industrialized societies and virtually nonexistent among our ancestors?

The first argument might be an objection that obesity has in fact been with us since the days of our earliest ancestors, so nothing has changed. It has been suggested that figurines of markedly obese women, found in Europe and dating to thirty thousand years ago, are proof that obesity existed at that time. However, no hunter-gatherer or small-scale horticultural group has ever manifested signs of obesity, despite having caloric intake and energy expenditure (adjusted for metabolism) within the range of contemporary industrialized populations. Thus the prehistoric statuettes may be representative of idealized feminine beauty, just as Barbie dolls and Japanese anime characters with huge eyes and exaggerated busts are fantasies more revealing of their creators than of real women.

Among the plumpest nonindustrial populations were the Inuit (the past tense is used in this paragraph because weight is changing dramatically among virtually all populations around the world). A male Inuit of Foxe Basin in eastern Canada measured on average 5'5" and 146 pounds, for a body mass index (BMI) of 25; women in the same group averaged 5' and 123 pounds, for a BMI of 24.[25] By comparison, the average BMI in the United States is around 29 for both men and women; obese is defined as a BMI 30 or over.[26] At the other end of the scale, Dobe !Kung males in Western Africa were a bit shorter (5'3") and a lot leaner (108 pounds), for a BMI of 19, while women were on average 4'11" and 90 pounds, for a BMI of 18. Since heat loss is reduced in bigger animals—bigger animals have more surface area to radiate heat, but their greater mass more than compensates for this loss—it makes sense that the Inuit and the !Kung, dwelling in the Arctic and sub-Saharan Africa, respectively, are at opposite ends of the BMI spectrum.

On the other hand, fat was likely considered desirable and sexually attractive in ancestral populations. In recent times, some groups in Africa and Oceania went to great lengths to fatten young people through force-feeding in preparation for marriage. However, the process was difficult, uncomfortable, and reserved for children of rich families who could afford fattening food and could do without the labor services of the participants. In societies where food was typically scarce, obesity was a sign of wealth, privilege, strength, and fertility rather than shame. To take one example, among the Massa of northern Cameroon and Chad, the lucky lad who was chosen for fattening went through a two-week gorging ritual. To make room in his stomach for the onslaught of food (twenty-nine pounds, in one observed case), he ate bitter roots that caused him to vomit and drank sour milk or other liquids that triggered diarrhea. He was then fed eleven meals a day of sorghum, milk, meat, and fat from six in the morning to four the next morning. The ordeal, marked by frequent vomiting, farting, pooping, and peeing, was considered painful and potentially hazardous. On the other hand, the fattened few were safe during the periods when most Massa suffered from greatly decreased food intake; the chosen were also considered the most sexually attractive.[27] Despite this grueling ritual, the young men typically lost their weight once the force-feeding ended. (However, they were believed to gain weight thereafter more readily than others.)

Though fattening among the Massa was reserved for privileged boys, the typical rule across most societies was that fattening for marriage was for daughters—and what glories their corpulence com-

manded! In recent decades, among the Annang of Nigeria, privileged
adolescent girls entered a fattening room and were fed copious quantities of food while refraining from any work. They slept on beds that
were deliberately uncomfortable, to keep the girls rolling about and
thereby softening their muscles, according to the belief. When the seclusion period concluded sometime between June and August, the
girls emerged to dance in front of the villagers on two occasions, once
in the village square and once in the market, naked save for bells about
the waist, blue beads in the hair, and heavy brass bracelets around
the legs, to show off their privileged rolls of fat. Among the neighboring Efik in Old Calabar Province, the daughters of rich families were
fattened on a deluge of food and freed from physical labor, after which
they were adorned with beads and bracelets and danced naked before family and townspeople "with an air at once arrogant and quer-
ulous."[28] As with the privileged young Massa men, gaining weight was
not easy; despite all the force-feeding, some young Annang women
were unable to gain much weight and remained moderately proportioned.

In view of the difficulty in traditional societies of attaining substantial weight, and the apparent similarities in terms of caloric intake and
physical activity levels between nonindustrial and industrialized populations, why has obesity risen so dramatically in industrialized
populations? One possibility is that the pattern of eating has changed
drastically, with adverse consequences for health. In nonindustrial
societies, as we discussed earlier, the number of calories people consumed varied over the seasons from feast to famine, while in industrialized societies, people's caloric intakes are close to constant. There has
been an explosion of interest in intermittent fasting diets, like the 5:2
diet (five days of regular eating and two days of fasting), because they
seem easier to comply with than regular diets, but scientific studies on
the effects of such diets are just starting to be undertaken. Small-scale
studies following rats and humans for weeks or months of alternate-
day fasting or intermittent fasting a few times per week seem mostly

promising so far, with modest improvements in weight, body fat, and heart and brain functioning, as well as improved risk factors for diabetes and heart and brain diseases.[29] (However, one study on rats found that the diabetes risk factors of abdominal fat and glucose intolerance worsened on an intermittent fasting diet compared to rats that ate freely or that were on a calorie-restricted diet.[30])

Religious fasts provide an opportunity to assess the health consequences of seasonal fasting. Greek Orthodox observe three major fasting periods, during the Nativity, Lent, and Assumption. Fasters avoid eating dairy products, eggs, and meat, as well as fish and olive oil. Outside of the three major fasting periods, followers abstain on every Wednesday and Friday, except during the weeks of Christmas, Easter, and Pentecost. All told, 180 to 200 days have dietary restrictions. (Nowadays, apps on iTunes are available to help Greek Orthodox followers remember the injunctions.) Studies on the effects of these fasts show modest health benefits, including lowered levels of LDL cholesterol.[31]

During Ramadan, the ninth lunar month of the Islamic Hijri calendar, healthy males are expected to abstain from food and drink from dawn to dusk.[32] Typically, one large meal is consumed after sunset, and a smaller meal is taken before sunrise, though some Muslims also eat again before sleeping. Studies of the health effects of the Ramadan fast show mixed results, which is not surprising considering the extensive geographical and cultural range of the more than one billion Ramadan followers.

Obesity, then, can be linked to changes in eating patterns, but the evidence for this is not very strong, so far. What about physical activity? As we noted previously, energy expenditure levels in industrialized societies are not that different from those in hunter-gatherer societies; moreover, since the 1980s, energy expenditure increased in the United States and Europe, just as obesity surged.[33] Critics of the link between physical activity and obesity have pointed out that exercise just makes people feel hungrier and eat more, and the body compensates by lowering metabolism, obliterating any gains from working out.[34]

A key factor may be physical *inactivity*. Prolonged stretches of watching TV, sitting down, and commuting by car have deleterious long-term health effects, including weight gain and diabetes. Our ancestors rarely spent prolonged periods being immobile; they couldn't, because that would have meant starvation, thirst, loneliness, etc. Instead, they were highly mobile; contemporary hunter-gatherers cover around 8.8 miles (men) or 5.9 miles (women) on foot each day.[35] By contrast, the average American walks about 2.5 miles each day. The things that are done instead of walking—watching TV, sitting at a desk, and driving—are all associated with obesity, disease, and early death. Watching television is associated with an increased risk of obesity, type 2 diabetes, cardiovascular disease, and outright dying. The average American now watches nearly five hours of television a day. Each two-hour increment in watching TV translates into a 23 percent increase in the risk of obesity, a 14 percent increase in the risk of diabetes, a 15 percent increased risk of cardiovascular disease, and a 13 percent increased risk of dying.[36]

Working while sitting, as in desk or computer jobs, had less drastic but still unfortunate consequences, with each two-hour increment being associated with a 5 percent increase in the likelihood of obesity and a 7 percent increase in diabetes. By contrast, one hour per day of brisk walking is associated with a decrease in the likelihood of obesity by 24 percent and diabetes by 34 percent.[37] Cars are also problematic. In an Australian study, people who commuted by car gained 4.8 pounds over four years, compared to 2.8 pounds for non-car-commuters.[38] Long-haul truck drivers have an especially tough problem with obesity: The obesity rate (BMI 30 and above) among them is 69 percent, compared to 31 percent among the general population.[39] The same problems afflict our pets: Dogs and cats that have to deal with living indoors, living in apartments, and being physically inactive also end up overweight and diabetic.

Now, some readers at this point will object: What is it about physical inactivity—watching TV, driving, and, for dogs and cats, being stuck all day indoors—that's so harmful? Didn't I just claim earlier that lowered energy expenditures are not to blame for the epidemic in obesity?

Several researchers have suggested that part of the reason that our pets are getting more obese is that their lives are boring; psychologists have also noted a connection between human obesity and boredom. Also, obesity rates are high in "boring" jobs that involve a lot of monotony (like driving trucks, cleaning buildings, factory work, and even construction jobs) and low in occupations that might seem sedentary but are intellectually stimulating (professors, teachers, artists). The usual claim is that the connection between boredom and obesity lies in stress, but stress is a tricky thing to define and measure. Who lives the more stressful life: a homeless person, a business executive with shareholders to placate, or a housewife burdened with social isolation and a philandering husband? Moreover, no matter how stress is defined, it has no consistent relationship with overeating.

Thus stress is not a helpful concept in understanding obesity. Rather, a key issue may be how energy is allocated within our bodies. Energy can be stored in fat cells, but it can also be used in powering the brain. People who have higher IQs and more education are at lower risk of becoming obese. It could be argued that highly educated people are less obese because they learn in college that low-fat foods are less fattening; however, groups like the Maasai traditionally ate a lot of fat in their diet but remained thin, and public perception about dietary fat and getting fat is not exactly a secret hidden in college campuses. Another argument about education and obesity is that the same willpower that gets people through college can also be applied to resisting tempting foods, but as pointed out earlier, obesity is not a problem of overeating; our lean hunter-gatherer ancestors ate about as much as we do today and expended about the same amount of energy.

An alternative explanation for the link between intellectual activity, boredom, and obesity is that the energy that could be channeled toward fat cells may instead be channeled toward fueling the brain, *if* the brain is occupied with challenging tasks.[40] When we engage in mind-stumping tasks, our glucose levels drop. In other words, "mental effort" is more than a metaphor; it takes energy to reason intensely. That's why glucose drinks and breakfasts improve mental functioning, even for dogs.[41]

As societies have become more peaceful and orderly over the decades and centuries, the day-to-day drama of finding food and a place to sleep at night and avoiding predators and disease has been eliminated and replaced by the safe but predictable routine of television, cars, computers, offices, factories, shopping malls, supermarkets, stationary bikes, and treadmills. Being free from the threat of assault and disease is a remarkable step forward for our species. However, the challenge of surviving in the wild has been replaced by the challenge of trying to stay awake amid maddening drudgery, and all of this drudgery may mean fewer calories getting used by our brains and more ending up stored in our fat cells. Indoor dogs and cats and not a few zoo animals might express the same regret, if they could only speak.

Other factors that may influence obesity include hormones, antibiotics, and birth control. It has been observed that pets that are neutered are at greater risk of putting on fat; with humans, men who have less testosterone and women who have less estrogen are at greater risk of becoming obese.[42] Children who are delivered via Cesarean section or who use antibiotics may have an increased chance of weight gain because of alterations in their gut-microbe community. (Although the exact mechanisms are still being worked out, farmers have routinely exploited the fattening effect of antibiotics to increase the weight of their domestic animals.[43]) The precise details regarding the links between obesity, hormones, and antibiotics are still being worked out.

This brings up a fascinating paradox. Several studies have found that people who are slightly overweight, with BMI from 25 to 30, tend to live longer than people who are considered to be normal weight (BMI from 18.5 to 25) or obese (BMI 30 or greater). Part of the reason may be that sick people and smokers tend to be thinner, but even after controlling for such possibilities, the overweight-longevity paradox remains.[44] There are at least two possibilities why being overweight may be healthier than being "normal" weight. Chronic diseases tend to result in loss of weight, muscle, and bone mineral density. Also, fat could help sequester and buffer the effects of toxins.[45]

Apart from the question of the ideal body type for long life or good health, there's another important consideration that influences many people: What's the ideal body type in the eyes of those we wish to attract? A surprising and perhaps disturbing conclusion from this research is that we tend to overshoot on our estimates of what it takes to look beautiful. According to studies of American college students, men want to be more muscular and bigger than women actually prefer; conversely, women want to be smaller, shorter, and more toned and have longer hair and bigger breasts than men actually prefer. What's going on? Why do we go through all this madness of trying to adjust our appearance if our partners are really not happy with the results?[46]

There are two possible explanations. The first possibility is that the important thing might not be the end objective; it might be more important for us to just have a strongly motivating goal in mind. If you are looking to attract a particular type of person, then having an exaggerated notion of the ideal body may be the simplest option available for achieving that goal.

A second, more likely explanation was offered by three of my colleagues at UCLA, David A. Frederick, Daniel M. T. Fessler, and Martie G. Haselton. They reasoned that a runaway competition for prestige takes place whenever we think about bodily characteristics and compare ourselves to others. We try to outdo others—that's human nature. Never mind what someone else wants; that's a pretty tough one to figure out, as any couple can attest. Just go one better than your peers, or imitate the most popular and richest public figure you admire, and you'll have an easy guide. Some Hollywood starlet wears fur boots and oversized shades and dyes her hair blond? Some Hollywood actor pulls off his shirt to reveal razor-cut abs? Got it. Such efforts might not be exactly what our partners want, but our minds are designed to make us compete against our peers in a silly but intrinsically human game of envy and status-seeking.[47]

The Greek island of Ikaria is famous for having some of the longest-living people in the world. Is this due to their diet or something else? To get a better sense of how Ikarians achieve their longevity, I catch a ferry from Athens to Ikaria, just off the shores of Turkey. My Ikarian friend George is an easygoing guy, quick to grin, always willing to chat, never in a hurry. He works at a grocery store. He wears no watch. During one visit to his store, George confides his wisdom to me, the key to Ikarian long life: "Great food. Great wine. Great sex." I ask him to repeat it. He obliges: "Great food. Great wine. Great sex." I heard him the first time, but he has an expression that I can't read, something between joy and absolute contentment—the ingredients to long life. After George and I run out of things to talk about—and after I get anxious about time spent away from my writing duties—George rings up my purchases: bread crisps, lentils, broad beans, chickpeas, cucumber. I take the road back to my hotel, a narrow lane hugging a cliff, the restless sea surging below.

Many people have visited Ikaria to learn about the secrets to long life. *National Geographic* has been here, a slew of scientists, Oprah's crew, all wanting to know how the people here live to such a ripe old age. The Ikarians I talk to dismiss the idea that their diet is the key to longevity. The real reason: no stress. The favorite saying on this island is "Don't worry." People talk slowly here. No rush, no worries. The magic works wonders on me. Every morning, I wake up beaming. Crisp air, bright sky, glimmering ocean.

One afternoon, I walk into a restaurant in Ikaria to return a glass bowl that had been lent to me. There are four ladies at a table, smoking and chatting. I often see them in the restaurant at this hour; by the dock, groups of men are also chatting, waving their hands like orchestra conductors, drinking coffee or the local brew, ouzo. The ladies ask me when I am leaving Ikaria. "Tomorrow," I reply.

The owner of the restaurant points to an "Ikarian Clock" on the wall. The clock has no hands.

All day, I have felt anxious about not getting work done, about not being able to make enough contacts or friends on the island, about money worries. I searched for flights and ferries out of Ikaria, ways of getting to a library, so I could push ahead with my schedule. After I see the Ikarian clock, though, something in me clicks. There is something extremely inviting about a land where time doesn't dictate your life.

I ask a young filmmaker on the island if she has eaten dinner. It is after 6:00 P.M. She laughs exuberantly, as if I just delivered the best punch line she has heard in ages. Dinner? She hasn't eaten lunch by that point. What's the rush?

Perhaps I won't be leaving the next day after all. Some things can wait . . . can't they?

If the Mediterranean diet is the gold standard of Western nutrition, then the Cretan diet is the pinnacle. The famed traditional Cretan diet was based on bread made from wheat and barley, tomatoes, dandelions and other mountain greens, cabbages, eggplant, okra, leeks, onions, radishes, olives, grapes, the "poor man's meats" (beans, peas, lentils, chestnuts, almonds, walnuts, and peanuts), some goat and lamb, fish, goat cheese, snails, and copious olive oil and red wine. Cretans had notably longer life spans than other citizens in the Mediterranean Basin. However, Cretans complained that their olive-oil-based diet left them hungry; 72 percent of families surveyed in 1948 mentioned meat as their favored dish.[48] During the following decades the Cretan people made good on their unfulfilled wishes. By 2010, the average middle-aged Cretan man ate around four times as much meat, almost twice as much pasta, and half the amount of olive oil and bread.[49] He was a spindly 139 pounds in 1965, but in 2010 he weighed around 183 pounds. (Cretan women are even more obese).[50] The rate of type 2 diabetes also spiked.[51] Paradoxically, Cretan men may eat *fewer* calories per day now than in the 1960s, due primarily to reduced use of calorie-rich olive oil.[52] If we wish to recommend the Cretan Mediterranean diet as the best, we should understand why Cretans were eager to abandon their traditional diet and why obesity rates were quick to jump upward despite the reduction in calories.

As a first guess, you might think that driving and the mechanization of farm work have cut back on the physical demands of Cretan farmers. After all, they used to burn more than 3,000 kcal per day on the job, but many Cretans now lead a sedentary lifestyle.[53] The reduction in physical activity could certainly be a major factor in burgeoning Cretan waistlines. However, Greek children living in rural areas are generally more obese than city kids, even though rural kids are also more physically active and more fit, by measures of running, jumping, throwing, and so on.[54] As previously mentioned, the key to obesity does not necessarily lie in energy intake or usage; rather, the answer may lie in physical inactivity, TV, cars, and boredom. Crete, in particular, went from being a place where walking and riding donkeys was the norm to something like a Grand Prix race circuit, with cars zooming madly from village to village, despite the compactness of the island.

I have lunch with a young family in one Cretan village. The father is a truck driver and busy that afternoon. The mother, pale and pretty, eyes darting nervously behind her spectacles, does her best to look after her three children, two boys and a girl. Lena serves us a lovely meal of slow-cooked goat, potatoes fried in olive oil, yogurt, and a salad bathed in olive oil. There is a telephone call; Lena goes out, waits for a school bus, collects her daughter, then holds her hand as cars careen by in either direction before they can cross the street back to their modern, spacious house. One of the boys gulps down his meal so that he can get to the part that he really wants to eat, the sweets. The other boy is extremely hyperactive and races outside as the mother calls to him to get inside and eat lunch. The girl is quiet and chubby and wears glasses.

"I want her to lose weight," the mother says wistfully. After lunch, all three children sit down to watch TV; first one show, then another. Lena is prediabetic. I ask her how much exercise she does every day.

"I have no time for exercise," she answers.

Given the glorious surroundings and the mild, sunny climate, I am surprised at first, but upon reflection, I realize it is the same answer that any of my friends with kids might give in California, which has a mild, dry climate like Crete's. When Lena drives me to the village, I suggest that we walk instead. She seems puzzled, then grateful for the chance to exercise—but it is less than a five-minute walk.

As a counterpoint to Crete, consider the Greek island of Hydra, situated about sixty-two miles southwest of Athens. Hydra has been a magnet for writers, artists, and musicians; Leonard Cohen was productive during his frequent stays there. Due to the close proximity of the hills and the steep terrain, the island's roads have never been favorable to motorized vehicles. Donkeys and getting about on foot are the best forms of transport here. In 1991, the Piraeus prefecture government formalized the status of Hydra as a car-free haven. The island now pulls in droves of tourists, and it's easy to understand why, once you wander along the maze of lanes, with nothing to overtake you except for plodding donkeys, no noise to disturb your peace except for church bells and cats. Hydra is scaled to human walking distances and speed. Complete the picture with the Mediterranean weather, the glistening waters, the ready access to the company of other people, and the bounty of red wine, and you can begin to understand why artists found more inspiration here than in the rush of city life.

Ironically, ever since Greece slipped into a debt crisis, the quality of life has improved in places like Athens. When I first arrived in Athens four years ago, at the start of the recession, the air was the color of pea soup and the streets were choked with cars, motorbikes, taxis. Now the streets are quieter. My Athenian friend and I hike to the top of a hill overlooking Athens. We are able to see far across the valleys, the hills sparkling with whitewashed homes, the cargo boats plying the ocean. After the recession began, many people left the city, returning to the countryside to look for work. People in Athens started using their cars less; some people took up cycling. The air quality showed improvement in concentrations of notorious acid rain components like nitric oxide and sulfur dioxide, and the ozone layer was replenished, as vehicle emissions decreased (though a new tax on oil use caused people to switch to burning wood, causing smog levels to increase).

When Greece's economy finally revives, people will likely go back to their old ways, ditching the bicycles and driving cars again, polluting the air and erasing any health gains attained from the temporary flurry of increased exercise. The reason Hydra is a car-free paradise has little to do with farsighted city planning and much to do with being located on hilly terrain on a small island. This is not to say that it

is impossible to ban cars; around the world, many communities, particularly in Europe, and particularly islands, forbid or greatly restrict
car traffic to make their streets safe and quiet. Once this is achieved,
the communities become attractive places to visit and live, and much
better for the waistline.

Another region renowned for its diet and health is Okinawa, an island group that is part of the Ryuku Arc in southern Japan. After reading about the miraculous food and health of the Okinawans, I am
eager to see the tropical islands with my own eyes and to sample famous dishes such as bitter melon, the zero-calorie root extract known
in English as konjac or devil's tongue, and pig's ears. At the first opportunity, I catch a plane from Sapporo to Naha, the Okinawa Prefecture capital. Back in 1949, Okinawans consumed on average
1,800 kcal per day, yet exertions from their mainly farming lifestyle
burned up around 2,000 kcal per day, resulting in a deficit of energy
and wiry proportions (the mean BMI was 21.2). Not only did they consume few calories; their protein intake of 1.4 ounces per day made up
just one-tenth of their energy intake. Most of this protein came in the
form of miso soup (fermented soybean paste mixed with dried fish,
kelp, or shiitake mushroom stock) and tofu. Some scientists believe that
the calorie-restricted traditional diet accounts for the robust health enjoyed by the Okinawans several decades later, when the average life
span stretched out to 83.8 years, a full year longer than people in mainland Japan, already the world's longest-lived country, and five years
longer than Americans. What made this achievement especially remarkable was that Okinawa was considered to be Japan's most impoverished, backward region.[55]

True to billing, the topaz waters around the islands are a delight
to snorkel, though I stupidly skimp on sunscreen, hoping that the melanin in my skin will shield me from sunburn. Dream on! The skin
on my back, after months of enduring blizzards and frost in Ottawa
and Sapporo, blisters like pork fat on a grill. As for the remarkable

age-defying Okinawan cuisine . . . I find instead sausages, Spam, eggs, and burgers, everything deep-fried, stomach-clingingly greasy. What happened?

In a plot twist the Marquis de Sade could not have devised with more cruel irony, the prefecture went from being Japan's healthiest to one of its sickliest in just a few decades. The "26 Shock," as locals call it, saw male life expectancy nose-dive from fourth place among Japanese prefectures in 1995 to twenty-sixth place five years later.[56]

To sketch out the dietary debacle of Okinawa, we need to go back to April 1, 1945, the day when 50,000 troops drawn from the XXIV U.S. Army Corps and the III Marine Amphibious Corps landed. After the Battle of Okinawa, or the "Typhoon of Steel," as locals refer to it, weary Okinawan survivors spent the first few months in internment camps, completely dependent upon American rations: Spam, biscuits, dried ice cream, powdered milk, Lucky Strikes, even military jackets for those who had no clothing left. The humaneness with which they were treated by the American occupiers was sometimes overshadowed by the brutal incompetence of the postwar administration of Okinawa. The islands became known as a dumping ground for unwanted bureaucrats (some twenty-two bureaucrats cycled through the top post during twenty-seven years of occupation) and unsavory soldiers. During one half-year crime spree in 1949, American soldiers perpetrated twenty-nine murders, eighteen rapes, sixteen robberies, and thirty-three assaults on the Okinawan populace.[57]

Okinawa became a Cold War pawn. The Japanese government and imperial household, eager to be rid of their American conquerors but also valuing the Americans as a counterbalance against the Russians, made secret overtures to hand the Okinawan islands over to the United States as a convenient location for American military bases. Okinawa fell under the trusteeship of the U.S. military (officially the U.S. Civil Administration of the Ryukyu Islands, or USCAR) and became a critical staging point for military conflicts in Asia. USCAR played up the cultural "Ryukyu" distinctiveness of the Okinawans (who have a different cultural history and language than Japanese mainlanders), hoping to drive a wedge between the locals and the

Japanese, and built lavish cultural friendship centers around the islands. Okinawans quickly became hooked on beef, coffee, fast food, cars, and other staples of the occupiers, tossing aside the elders' mainstays of sweet potatoes and seaweed as famine food. As much as the Americans were an oppressive and much-hated presence on the island, native Okinawans couldn't help but adopt the American diet and lifestyle.

One night in Naha, after searching in vain for wholesome traditional Okinawan food, I end up in an open-air bar just off the main drag, where a barmaid shakes cocktails under a strip of electric blue lighting and brays at customers' jokes. I chat with a middle-aged Ryukyu man with puffy bags sagging under his eyes. "We don't like Japanese. We don't like Americans," he mutters drunkenly.

Trapped in the crossfire of empires, politics, and war, Okinawans have much to be resentful about. This time around, however, the enemy is not a soldier in a uniform wielding a rifle or bayonet but the burger, fries, and soft drink wielded by a pimply teen in a different uniform, as well as the car used to get to the fast-food joint. The evil rotting the prefecture's health comes in the form of cheap and addictive processed foods, the availability of motorized vehicles, the convenience of one-stop shopping at a supermarket or mall instead of visiting scattered stores, the television that obliterates social life, and the steady smoking habits of its citizens. As a result, Okinawans have experienced a surge in lung cancer rates, type 2 diabetes, waistline size, and suicides, and the aforementioned drop in life expectancy. As one doctor in Okinawa has put it, this is the second Battle of Okinawa, fought behind cultural lines, with perhaps just as many lives at stake.[58]

This chapter has so far dealt with calorie intake and consumption. However, we should not neglect the topic of what happens to food as it exits our body (or refuses to exit, in the case of constipation sufferers, who comprise some 15 percent of people in North America[59]). Throughout our evolutionary history, squatting was the normal way of having a bowel movement. Sitting toilets, which became common in Western nations from the nineteenth century onward, create an un-natural 90-degree bend for the passage of fecal material, and thus straining is required to expel feces. When we squat, this angle is straightened out completely, and therefore much less time (around one minute to complete a bowel movement from a squat versus two min-utes on a sitting toilet, according to one study) and effort are required to expel feces. This might explain why constipation, hemorrhoids, and diverticulosis (a condition causing pouches to form in the colon) are much more common in Western populations than in Asia and Africa, where squat toilets are the norm.[60]

Online stores sell kits to convert sitting toilets to squat toilets, or you can build your own convertible squat/sitting toilet using online examples as guides. Be forewarned: If you didn't grow up using squat toilets, it takes practice to learn how to poop from a squat. At least one study has found that there may be increased risk of a stroke from shifting to a squat-style toilet, due to the effort involved in squatting and standing, so the elderly and people with high blood pressure should consult with a physician before switching to squat toilets.[61]

# THE FUTURE OF FOOD

When the hype is greater than the science it burns all of us.

—K. Lance Gould, *quoted in Shari Roan,*
*"A Slow Change of Heart"*

M ost people today agree that something is wrong with our life-style habits, but there is sharp disagreement among both experts and the public on what needs to be done to restore our health. Are today's food and health activists on the right track or misguided? I met with three of the leading proponents of nutritional advice—Dr. Dean Ornish, a cardiologist and proponent of low-fat diets; Sally Fallon Morell, a booster of traditional American farm diets; and Mark Sisson, an ex-Ironman athlete and a blogger and writer on the Primal (Paleo) lifestyle—to see why smart people reached opposing perspectives on the optimal diet. I also visited food idealists in Australia, Canada, and the United States who, with courage and determination, are trying to change the way we live and eat or raise food, particularly foods that are more ecologically sustainable; happily, it turns out that eco-friendly foods are also more suitable for our nutritional needs. As we shall see in this chapter, however, these activists are being confronted with major hurdles due to the nature of capitalism and our fear of novel foods.

Dean Ornish is a busy man. Besides teaching medicine at the prestigious medical school at the University of California, San Francisco, Dr. Ornish was appointed by President Barack Obama to the White House Advisory Group on Prevention, Health Promotion, and Integrative and Public Health. He had previously been appointed to the White House Commission on Complementary and Alternative Medicine Policy by President Bill Clinton, and has served as "a physician consultant" to President Clinton since 1993 and to several members of Congress. Through his numerous books and articles, Dr. Ornish advises the public to eat less fat (preferably around 10 percent of their total daily calories), stay away from saturated fat and cholesterol, eat little meat, limit alcohol consumption, and eat a lot of whole grains. (In his other suggestions, such as eating vegetables and fruits, staying away from processed foods and sugar, spending time with loved ones, and exercising, Dr. Ornish is in agreement with most other food and health writers.) In several studies that he led, there is evidence that his low-fat/low-meat diets, in combination with moderate exercise, stress management, cessation of smoking, and group psychological therapy, lower the risk of heart disease without the use of lipid-reducing drugs.

This brings up three important questions. First, is it the low-fat and low-meat diet that does the trick in reducing heart disease, or are the lifestyle interventions of exercise, stress reduction, smoking cessation, and group therapy the real key? In conversation with Dr. Ornish, I point out the "Spanish paradox": Spanish people consumed 30 percent more fat in 1980 than they had in 1966, particularly saturated fat (a 48 percent increase), yet heart disease decreased over the same period; in Japan, there was a similar postwar increase in fat and cholesterol consumption while heart disease also dropped. Dr. Ornish responds, "One has to be very careful about drawing conclusions from just looking at one factor in a population when there are so many other things that have been changing during that time as well." Fair enough—the authors of the "Spanish paradox" study themselves were skeptical that increased consumption of fat in Spain led to decreased heart disease, suggesting instead that increased antioxidants in the Spanish diet from eating more fruits could have been the real reason behind lowered rates of heart disease (wine and sugar consumption dropped slightly). Conversely, perhaps the belief that saturated fat is artery-clogging was so deeply ingrained in the minds of the study's authors that the possibility of saturated fat lowering heart disease was literally unthinkable and heretical to them. Increased fat in the diet may replace foods more dangerous for heart health; for example, carbohydrate-rich diets (the Spanish diet in the 1960s was based heavily on bread, potatoes, pulses, and rice) raise serum triglyceride and VLDL (very-low-density lipoprotein) cholesterol levels, both major factors for heart disease.[1]

The second important question to ask: Does a low-fat, low-meat diet reduce overall mortality? After all, it's not very encouraging if we avoid the risk of heart disease but increase our risk of dying from something else. When I showed a video of Dr. Ornish to one of my friends in Los Angeles, she said, "Oh, he looks healthy. He's the same age as you?" She scrutinized the screen. "He has a lot of hair." I'm forty-one. Dean Ornish is *twenty* years older than me. So perhaps he has figured out the secret to long, healthy life. In some respects, Dr. Ornish's suggested diet looks similar to the diets of the peoples with the greatest longevity (Okinawans, Nicoyans in Costa Rica, Sardinians, Ikarians) and indeed of most people in the world prior to industrialization. It's low in meat—our preagricultural ancestors did a pretty good job of hunting out the big mammals, and climate change mopped up the rest—and high in plant foods. Also, as discussed earlier, the link between low protein intake and longer life has been studied and demonstrated in various animals, and it's likely that the human life span may benefit from low protein (especially low animal protein) intake as well.

This link between protein restriction and longevity is also consistent with evolutionary biology: Nature favors longer life spans in animals that don't have adequate nutrition to compete and reproduce at an earlier stage in life. To put it another way, eating meat and fat may help you be fertile, attractive, and strong at a younger age, but it will also help you into the grave a bit faster.

When I suggested the evolutionary biology scenario to Dr. Ornish, he was not fond of this interpretation—"I'm not sure that natural selection explains everything," he says. He believes that his diet promotes health at all stages of life, young and old, but I doubt any sumo wrestler or weight lifter would win a title on Dr. Ornish's low-fat, low-animal-protein diet. Girls who eat a lot of meat and dairy tend to reach menarche (their first menstrual bleeding) at an earlier age, and girls who have early menarche tend to die younger; women who lack body fat are more likely to be infertile.[2]

Additionally, a diet that promotes a long life span in youth may not necessarily be an effective diet for an elderly or sick person. The major health risks for a younger person stem from chronic diseases like cancers and heart disease, which develop over the course of decades. For an elderly person, by contrast, the important task is to weather any illness that strikes, in which case eating more animal protein may promote longevity. Also, people who are moderately overweight tend to live longer, as previously discussed, and one possible reason is that metabolic reserves may enable sick people to overcome disease.

The third question to ask with regard to a low-meat, low-fat diet: Is it easy to follow? A typical set of meals for Ornish and his family might include whole-grain cereal with soy milk and fresh fruit, whole-wheat toast, and pomegranate or orange juice, along with a cup of tea or coffee; alternatively, he might fix an egg-white omelet with spinach and mushrooms, or low-fat cheese or turmeric (reputed to have anti-inflammatory benefits). As a treat, Ornish and his family indulge in whole-grain pancakes or waffles with a little maple syrup. The family takes daily multivitamin and fish oil supplements. For dinner, they serve vegetables like corn, broccoli, and cauliflower (cooked in a steamer to preserve most of the flavor and nutrients), along with a few prawns or some fish. He argues that foods can taste delicious without adding a lot of fat, salt, and sugar, which he believes mask the true flavors of food.[3] Some health professionals bemoan the difficulty in getting patients to comply with Dr. Ornish's relatively bland, low-fat, low-salt, low-sugar fare. Still, it seems reassuringly familiar, the kind of food that most doctors and nutritionists today would recommend.

———

At the opposite end of the dietary spectrum are the nutritional activists who advocate a high-fat, high-meat diet. On a shimmering blue morning, a woman with a tidy press of white curls greets me at her farm. Sally Fallon Morell is the co-owner of and the force behind P. A. Bowen Farmstead, a sixty-acre farm an hour's drive from Washington, D.C. Sally proposes that we tour the cheese-making operation housed inside. We don white coats and hairnets, dip our plastic shoes into antiseptic pools, and stroll through sparklingly clean rooms holding racks of cheddar and blue cheese in various stages of processing. Leaving the building, we walk along dusty roads to the pasture area, where chickens roam in grass recently vacated by cattle and feed on maggots that have emerged from the cattle dung. We pass a fish pond stocked with bass, sunfish, catfish, koi, and minnows, along with a handsome flock of sturdy Silver Appleyard ducks, then trudge over to a patch of forest where a herd of hogs, a heritage breed of Berkshire, Tamworth, and Spotted Pig, grunt excitedly in response to Sally's shrill call. Besides one day providing meat, their job is to clear undergrowth, eventually rendering the forests suitable for the cattle to range through. When we amble over to a herd of grazing cattle, Sally pauses to enjoy the sight.

I've been walking around in a button-up shirt and stiff black dress trousers under a hot sun, so I'm relieved when we return to the main farm building. Today is the weekly chicken slaughter day. Under the shade of a roof, a team of men and women, young and old, work on a mini assembly line, slitting, bleeding out, scalding, and gutting a pile of chickens as country music pipes in the background.

When Sally decided to throw her full effort into this farm, some people raised eyebrows—after all, she grew up in an affluent Los Angeles suburb and is in her midsixties; her husband, a farmer from New Zealand, is eighty-eight, though vigorous. The farm is still in its early stages and has yet to turn a profit—but how can it? She feeds the chickens grass peas from Pennsylvania instead of cheaper genetically modified soy; she doesn't use antibiotics on her cows; she doesn't pasteurize her milk; her animals are all free-range. It's a noble effort, but she admits to "sleepless nights." As she says, "I'm the one who wakes up in the middle of the night. I have tremendous sympathy for farmers." Her husband helps with tractor work, but Sally is the one who pours her money and soul into the farm.

Sally Fallon Morell is a fighter in another way; she's perhaps the most controversial nutritional activist in the United States today. The bestselling book that made her famous is provocatively titled *Nourishing Traditions: The Cookbook That Challenges Politically Correct Nutrition and the Diet Dictocrats* (written with Mary Enig, a nutritionist and biochemist, and first published in 1995). The introduction ranks among the most rousing and incendiary calls to arms ever penned in a cookbook. Sally and Mary take aim at the nefarious "Diet Dictocrats," whose sundry ranks include "doctors, researchers, and spokesmen for various government and quasi-government agencies," such as the Food and Drug Administration, the American Medical Association, the National Institutes of Health, medical schools and nutrition departments, and the American Cancer Society and the American Heart Association, which are "ostensibly dedicated to combating our most serious diseases."[4]

Sally and Mary's book made waves by insisting that a healthy diet includes a lot of fat (including saturated fats), cholesterol, salt, calcium, raw milk, and fermented foods, and few to no soy products; basically, it's a traditional American farm diet. In the book, Sally approves of the "five *B*'s": bacon, butter, beef, sourdough bread, and blue cheese.[5] (When I meet Sally on her farm, she points out that beans are also praiseworthy.) Further fanning the flames of controversy, Sally and her colleagues offer legal assistance to farmers who sell raw milk—illegal in parts of the United States and all of Canada and Australia, but legal in most of Europe—and argue for the merits of a raw-milk formula over breastfeeding for women who are deemed unsuitable for breastfeeding.[6]

Though one might naïvely suppose that Paleo/low-carb practitioners would be close allies in the promotion of animal-rich fare, Sally has castigated the Paleo diet for excluding agricultural products like grains, beans, and dairy, and for being too miserly with allotments of fat and salt. As she wrote on her foundation's Web site, "What does it do to the psychology of a growing child . . . to deny them ice cream (homemade, of course), whole milk, sourdough bread with butter, baked beans, and potatoes with sour cream?" She contends that children "need to grow up on a diet that says, 'Yes, you may,' not 'No, you can't.'"[7] The passage is a beautiful example of the impassioned and intuitive arguments that have made Sally and her food activist organization, the Weston A. Price Foundation, a force to be reckoned with in the nutrition wars convulsing the American dietary landscape.

One aspect that has drawn the ire of traditional and Paleo followers is that Sally and her coterie have claimed the mantle of Weston A. Price. Price was a Canadian-born dentist who practiced in North Dakota and Ohio. Beginning in 1931, accompanied by his wife, he examined the link between dental health and food among various groups around the world, including the Swiss, Celtic fishermen, South Pacific Islanders, African tribes, and native groups throughout the Americas. Weston A. Price concluded that the introduction of Western processed foods, particularly sugar and flour, hastened the development of cavities. Conversely, a diet consisting largely of traditional food, from fish to moose to coconuts, protected against cavity formation and promoted good physical health overall.

To go from the conclusion that almost all traditional diets free of white flour and sugar were healthy to the assertion that a healthy diet is only one that includes relatively high levels of fat, cholesterol, and calcium is an interpretation of Price's observations. By comparison, another organization that has adopted Price's name, the Price-Pottenger Nutrition Foundation (formerly known as the Weston A. Price Memorial Foundation), makes the case that healthy traditional diets include "minerals and fat-soluble vitamins found in butter, sea foods, fish oils and fatty animal organs," "raw, unaltered proteins from meats, sea foods, nuts, raw dairy and sprouted seeds," and "sweeteners rarely and sparingly."[8] This is a milder, more inclusive set of criteria than the Weston A. Price Foundation's, but how many Americans have heard of the Price-Pottenger Nutrition Foundation? With such a blasé stance and no engagement in mudslinging, PPNF is like a prim wallflower at a college party, destined for obscurity. Moreover, a diet of bacon, butter, beef, sourdough bread, and blue cheese resonates with more Americans than, say, a traditional Okinawan diet of sweet potatoes, bitter melon and copious greens, fish, soybean products, and small amounts of pork, never mind the science supporting the Okinawan diet.

Weston A. Price, the dentist, was correct in surmising that traditional diets from any part of the world were effective in protecting the health of the eaters. The specific traditional American farm diet proposed by Sally—the six *B*'s of bacon, butter, beef, sourdough bread, blue cheese, and beans—would have been suitable under the tough, vigorous working conditions of early American farmers. One only needs to glance at photos of early farmers to see that these were lean, healthy people, unfamiliar with obesity and its related diseases. For sedentary Americans today, a low-fat, low-meat diet like that proposed by Dr. Ornish is more likely to be healthy with respect to chronic diseases, because his diet is less energy-dense—little sugar or fat—and has less animal protein, but sustaining the willpower to deny oneself delicious fatty, salty, and sweet fare can be a monumental feat. The binge eating or compulsive snacking that results when willpower breaks down can undo all the possible advantages of a low-fat, low-meat diet. The better route, as I will explain, is to shift our lifestyles so that more moderate exercise is included in our daily routines, allowing us to eat with less guilt or fear of harmful consequences.

Mark Sisson bounces into the Malibu café looking like a movie star: loose T-shirt, long wavy white hair, well tanned. He orders an omelet with avocado, bacon, chicken, feta cheese, mushrooms, and onions. The dish is served with potatoes, but Mark mostly ignores these. A former long-distance runner, triathlete, and Ironman competitor, Mark is the author of *The Primal Blueprint,* a guide to a Paleo lifestyle, and maintains an influential blog on Paleo matters. Like Sally, Mark eats a lot of fat (50–60 percent of his calories) and likes his dairy, but unlike her, Mark bemoans America's addiction to simple carbs: "I think of potatoes much the same way that I regard most grains, which is it's sort of a beige food that's a source of cheap calories that convert to glucose pretty quickly." He regards his plate. "I don't include potatoes in my eating strategy. But mostly because I'm just not impressed with the way they taste. You have to put on a whole bunch of stuff on them to make them taste good." Mark goes on to recount his disdain for agricultural staples like wheat and oatmeal, which also require much flavoring to make palatable and are too easily converted into glucose.

Based on his reading of the ancestral human literature, Mark figures that it's okay to eat when you're hungry and abstain when you're not, instead of following a regular three-square-meals-a-day plan, and that exercise should consist of a lot of moving about rather than strenuous workouts, which Mark believes caused him to be frequently sick and injured during his competitive days. His version of Paleo puts an emphasis on lifestyle and, unlike pure-form Paleo, includes dairy.

Mark asserts that after cutting grains out of his diet, he cured himself of lingering arthritis, irritable bowel syndrome, colds, persistent sinus infections, and heartburn. The key, Mark suggests in his book and on his blog, Mark's Daily Apple, is the role of insulin. When too much simple sugar enters the body, it leads to loss of insulin response, which in turn floods the body with glucose; this excess glucose interferes with protein function by sticking to the proteins and creating advanced glycation end products (AGEs), which are believed to accelerate aging processes, including chronic inflammation.

One potential complication with the Primal/Paleo philosophy is that some of the longest-lived groups on the planet historically ate high-carbohydrate, low-protein diets. The Okinawans, as previously described, subsisted largely on vegetables, sweet potatoes (introduced from Central or South America via China in 1606), rice, tofu, fish, and sake (rice wine); whale blubber, when it could be obtained, was eagerly consumed.[9] The long-lived Costa Rican Nicoyans dined on corn tortillas cooked with lime, rice and beans fried in pork fat, boiled plantains, bits of meat and fat, fried eggs, vegetables, and large quantities of tropical fruits.[10] On Sardinia, a centenarian bastion off the west coast of Italy, as late as 1941, a typical day's meal offered a kilogram of bread, an onion, some fennel or radishes, beans, perhaps goat's milk or mastic oil, minestrone soup in the evening, and not more than a quarter bottle of red wine. Richer folks would add cheese or pasta to this diet.[11] Moreover, as previously discussed, caloric restriction studies seem to indicate that protein restriction is as important as or more important than calorie restriction in promoting long life. When I ask Mark about the long lives of people in places like Okinawa and Costa Rica, he replies that people there were happy by nature and dealt more effectively with stress. Mark also points out that these people were highly physically active and did not have continuous access to large quantities of food.

A frequent claim by Paleo followers is that they lose weight and eventually feel better on low-carb diets. However, the same claims of weight loss and mood improvement are also made by people who maintain vegetarian or raw food diets. There is some evidence that in the initial phases, people do lose weight more quickly on low-carb diets than on conventional low-fat weight-loss diets, with no short-term adverse health consequences. However, it seems that this weight is gained back in the long term. Additionally, eating a lot of meat may shorten life span, as previously mentioned, particularly for people under the age of sixty-five, when the risks of chronic diseases stemming from meat or fat consumption are of concern. On the other hand, also previously noted, people over the age of sixty-five who consume a lot of meat may live longer, because chronic diseases tend to take a long time to develop, and the health concerns of older people are linked instead to issues like frailty and wasting.

While it's hard to provide a comprehensive theory that will cover everyone at every stage of life, it's likely that low-carb diets are most harmful to children and most beneficial to older adults. For middle-aged people, consumption of cholesterol and fat is likely to improve mood and sex drive, while there is not much evidence for long-term weight loss. A better road to weight loss than a radical low-carb diet is to change our lifestyles so that we get more moderate exercise, which Mark himself advocates. It should be kept in mind that people whose traditional diets were meat-heavy, such as Arctic peoples, may do best by continuing to eat such diets, given the complications of genes that are not adapted to high-carbohydrate or high-calcium diets.

Back to our original question: Why do smart people disagree vehemently on something so basic as a healthy diet? One reason for the disagreement is that each of these diets has different health effects, both good and bad. Dean Ornish's low-meat, low-fat diet is the best bet for a long life, due to the effects of cutting back on animal protein, but it is the least psychologically satisfying and therefore also hard to maintain. Sally Fallon Morell's and Mark Sisson's high-dairy, high-meat diets are more likely to lead to shorter lives, but such foods are also likely to make people feel better and more likely to improve muscle mass. Moreover, high animal protein diets may be beneficial for the elderly, due to the problems of frailty and wasting associated with advanced age.

At a deeper level, the approaches of nutritional activists like Ornish, Morell, and Sisson conflict because they analyze food in terms of nutrients—protein, fat, carbohydrates, sugar, vitamins, etc.—rather than adopting a nuanced view of evolution. Breaking food down into its nutritional components has brought many advantages, such as the elimination of diseases like beriberi, pellagra, and rickets, because it allowed scientists to determine which nutrients were missing in modern industrialized diets and lifestyles. However, due to the complexity of human physiology, the ethical barriers to human experimentation, and the great variability of our ancestral diet and our gene pool, the monumental scientific efforts and funds poured into nutritional research have yielded disappointingly scant progress since the conquering of beriberi, pellagra, and rickets in the first half of the twentieth century. This leaves the public understandably bewildered and frustrated about what to eat to maintain or restore health.

A major flaw in nutritional research has been neglecting the insights that evolutionary theory offers. Without understanding the evolutionary history behind humans, trying to determine the optimal diet is like trying to decipher a difficult text by reading only one page;

only evolutionary theory provides the means to understand how all the components of an organism's life are linked together, including nutrition and health. On the other hand, overly simplistic interpretations of evolution, such as viewing the human ancestral diet as consisting primarily of meat, also deprive us of valuable insights into nutrition and health.

When we put a nuanced view of evolutionary theory back into nutrition and health, we end up with the following observations:

- If we do not exercise or make an attempt to be physically active, then we are much more vulnerable to chronic diseases, regardless of food choices; conversely, if we exercise or are sufficiently physically active, then we can avoid chronic diseases while eating liberally. This is because humans evolved in a context of constant movement and moderate physical activity; sitting for prolonged periods was extremely rare, because that would have led to starvation or the loss of opportunities to socialize and, therefore, reproduce.

- To get a balance of nutrients, we should eat traditional cuisines, the older the better (for example, from five hundred years ago), because traditional cuisines were carefully pieced together through trial and error. Focusing on nutrients is often a fool's errand. For example, eating less meat and fat can be harmful if we end up craving sugary foods instead. Traditional cuisines get around this problem by offering balanced, tasty meals. For people who trace their ancestry to a specific region of the world, the traditional cuisine from that area is likely best suited for their genes.

- Eating a lot of animal foods when you are younger will make you grow taller and stronger and be more fertile and attractive but will increase your risk of dying earlier. As we discussed, this trade-off between robustness in early life and poorer longer-term health is exactly what we should expect from an evolutionary perspective, because evolution is concerned only with the passing on of genes to the next generation, at whatever cost necessary to the current generation—poorer long-term health being such a cost.

There are other aspects of food and cooking, besides nutrients, that are critical to our well-being. For example, the fact that meals containing little meat may be healthier in the long run may not matter much if people who are poor or lack knowledge of cooking are unable to prepare savory low-meat meals. How can we ensure equitable access to healthy cuisine, and how do we make these ingredients sustainable for a larger population? The city of Melbourne, I discover, is a good place to examine trends in food fairness and sustainability.

Melbourne has recently been the site of an unusual food innovation: pay-what-you-can restaurants. Fork out a buck or two, or cough up a hundred, it's all up to you; you'll still get the same meal at one of three restaurants called Lentil as Anything, a reference to an Australian art-school new wave band, Mental as Anything. The name of the restaurant sums up the ethos of these eateries: a little bit crazy, a little happy, a vegetarian-anarchist wonderland.

A friend introduces me to the restaurant, taking me down a flight of stairs that spill onto a riverside path just a few hundred yards from the street. We step into precolonial Australia, dusty and dry; a path runs along a sluggish river, the banks crowded with eucalyptus. We follow the winding path, cross some sheep pens, and slip into the sprawling grounds of a former nunnery adorned with a spacious flower garden.

This is the flagship Lentil as Anything restaurant, pulling in a smorgasbord of gaunt artists and students, well-to-do liberals, and curious tourists. After stuffing all my spare change into the collection box—no one is looking, but my lawyer friend seems generous with her dollars—I stack my plate with South Asian–inspired cuisine: curries, fried pastries, rice, and coconut. I could eat much more of this delicious food, but would there be enough for the latecomers? Communal eating has this peculiar effect of forcing us to think of others' needs.

My friend and I eat indoors; the heat outside is sticky. Most of the eating space is taken up by long wooden tables that encourage mingling. The staff is young and diverse; the restaurant has a policy of helping to sponsor refugee applicants. The notion of eating the same good food as everyone around me is moving and inspiring. The fare isn't Michelin, but it is much better than what I can cook on my own. What made this generosity and camaraderie possible? And why aren't places like Lentil as Anything more widespread?

A week later, I meet with the founder of the restaurant chain. Shanaka Fernando is courteous and eloquent, speaking in a meditative, ruminating manner, like a monk or poet. His father was a Sri Lankan army officer, his mother an Irish potter who was disowned by her family for marrying a dark-skinned native. Shanaka grew up privileged in Sri Lanka, white beneath the dark skin: servants, security, the best schools. He lived through periodic violence, when the Sinhalese majority viciously persecuted Tamil minorities,[12] then came to Melbourne to study law to appease his father. He lost interest in his studies, dropped out, and opened a café. With characteristic impulsiveness and idealism, he wiped the prices off the board.

Why pay-as-you-can? Shanaka observed during travels among rural groups in Indonesia, the Philippines, and the Amazon that food was shared among neighbors. "Food is a strong gesture of our kinship. Whereas, I find that sometimes in Western societies, at least in Melbourne for example, whenever you go out to a restaurant to eat lobster, it is a means of highlighting privilege and separating yourself from a majority of society. I wanted to see how we could capture that culture of making food available and then seeing everyone eat together from all walks of life, especially because money is such a divisive force in society. I was curious to see if we could use the money that people donated or made available to unite people, the focus being the importance of having a good meal and being able to sit with the rest of the community, the rich and poor, everyone, together and eating."

Against the odds (and the objections of his then-partner), Shanaka opened first one, then two more pay-as-you-can restaurants, and a school canteen, drawing out of his skeptical Melbourne residents previously unthinkable fonts of generosity and trust. Shanaka was feted as an Australian Local Hero, appeared on a national stamp, met the prime minister, gave TED Talks, and co-officiated a TV cooking contest with the Dalai Lama. Shanaka's pay-as-you-can philosophy, considered subversive a decade ago, is now mentioned in Australian educational curricula. The idea has been exported to Dublin, and in 2011 Jon Bon Jovi opened a similarly themed community kitchen in New Jersey.

However, Shanaka paid a price for following his idiosyncratic path. Heroin addicts dipped into the collection boxes, which eventually had to be locked. He battled for years with the Australian government over $300,000 in unpaid sales taxes owed by Lentil as Anything; eventually Shanaka and his supporters succeeded in having the sales tax law revised, in view of the nonprofit nature of the restaurants. He had to declare bankruptcy at one point but formed another legal entity and bought back the restaurants' equipment. Shanaka pays himself a basic wage out of the earnings, though child support payments became a legal issue, and he has been threatened with jail over $14,000 in unpaid traffic fines. Businesses and landlords have pushed for the eviction of Lentil as Anything restaurants, in part because they draw in an unsavory crowd of social outcasts. Shanaka is weary of the battles and the restaurant business; he wants out. He has a lot of plans; he has worked with children's education and reconciliation in Sri Lanka. Like many of us, he wants to challenge deep-rooted social inequities; unlike the majority of us, Shanaka has the courage to do so.

Shanaka's restaurants prove that healthy eating does not have to be the privilege of the well-off in society. As Shanaka and many anthropologists have noted, in small-scale traditional societies, food was shared among neighbors. Indeed, the act of sharing food was essential to village life, because it meant that the risk of not obtaining enough food in a foray could be spread among the villagers. Nowadays, this communal aspect of eating has been nearly wiped out in industrialized societies, with people dashing off to a supermarket, farmer's market, bakery, or deli and returning home to consume the stash of food by themselves, perhaps with some family, occasionally with friends. Eating in a restaurant is not much different, because, as Shanaka noted, the costs of restaurant meals can serve as markers of status, like a fancy car, watch, or purse. For most Americans, this is exactly the point: If you work hard to make money, then you get to splurge on luxuries, including pricey meals. But this raises a question: Is a city or town merely a place with good jobs, safe housing, decent education for the kids, and places to blow excess cash for amusement, like malls and restaurants? Or are communities supposed to be places where we look out for one another, buffering our fellow citizens against the vagaries of misfortune? This might sound like a philosophical question to pose in a book about food and health, but in historical terms, the sharing of food and risk in general was the cornerstone of communal life.

It might be the case that Shanaka's idealistic pay-as-you-can vegetarian restaurants (plant foods can be easily donated as surplus, but meat is trickier) will eventually fail from too much freeloading; in any case, his venture forces us to consider a critical issue: Should eating be a private, self-centered affair, or can food be recast in its original role, as a means of binding and protecting citizens?

There are two ways in which an eater can act in a benevolent manner. The first is to assist fellow citizens, which is evident in Shanaka's pay-as-you-can restaurants, where better-off diners subsidize healthy food for the less fortunate. The other path to benevolence is to eat in a manner that safeguards the prospects of future generations. When we buy cheap meat, fish, and produce from a supermarket today, we are essentially being subsidized by future generations, who will have to pay more for the same meat, fish, and produce (if it can even be found), because there will be less fossil fuel and fish available, and because the planet will become progressively degraded by agricultural and waste-disposal practices that are oriented toward short-term profit and convenience. Food idealists are attempting to minimize the costs we inflict on future generations by eating and raising food in more ecologically sustainable ways. For example, instead of rearing imported animals and plants that harm local environments, consumers can opt instead for animals and plants that are well integrated into local ecosystems, as indigenous peoples necessarily did before the advent of global trade.

Consider this irony, then: In Melbourne, you can indulge in a panoply of cuisines—Italian, Japanese, East African, Lebanese, Moroccan, Vietnamese, Indian, and more, reflecting the diversity of the immigrants in the city—with one conspicuous omission: There is hardly any native Australian cuisine to be found. A government-funded restaurant called Charcoal Lane is a striking exception. Occupying a two-hundred-year-old building that once housed an Aboriginal health center, Charcoal Lane offers native Australian food and trains Aboriginal and non-Aboriginal apprentices in the kitchen and dining area. The manager of the restaurant, a Sri Lankan Aussie named Ashan Abeykoon, and the head cook, a white Australian named Greg Hampton, are doing their best to expose Australians to the plants and animals that thrive in the wilds of their country. My meal there—camel sausage, mutton bird (a seabird), salad of bunyon nut, wattle seed, and kumquat berry—was terrific, teasing the tongue with new flavors and sensations; I especially liked the fishy taste of the mutton bird, and the camel was succulent. Other offerings on the menu include wallaby, emu, and saltbush lamb.

Greg, who has been cooking for twenty-six years and at one point ran his own zoo, points out the environmental benefits of raising indigenous or desert-acclimated animals. When European settlers first arrived in Australia, they brought sheep, cattle, and pigs and cut down trees to grow wheat. Over time, the heavy use of water to grow wheat exacerbated the soil's salinity; as the water percolated down, it served as a conduit for minerals and leached them from the soil. The sharp hooves of the imported animals compacted the earth, destroying landscapes and increasing water pollution and sediment loss from runoff. The decimation of vegetation on riverbanks increased river flow and worsened the problem of soil erosion.

Greg points out that by contrast, native Australian plants have long roots, which allow them to tap water from deep within the soil without increasing salinity. Kangaroos, wallabies, and emus have relatively soft feet, which do not compact the soil, and they feed on native plants with long roots. The plants along riverbanks are preserved, and the currents of rivers are slower. Salt-adapted trees have small, intensely flavored fruits that contain high levels of antioxidants, including vitamin C. Kangaroo flesh has high levels of zinc (which plays an important role in the immune system). Saltbush lambs are not native, but because they are adapted to arid environments, they consume natural vegetation that is rich in salt and other minerals.

So with all the positive environmental and health benefits and superior tastes of native plants and animals, why aren't people flocking to places like Charcoal Lane? Ashan, Charcoal Lane's manager, explains that when the restaurant tried to offer "kangaroo tail" on its menu, people stayed away from it. The tail was unfamiliar to diners, and its location far down on the animal made the meat less appealing. If it was called simply "kangaroo," there was more interest. Still, kangaroo is unlikely to become a local staple. The kangaroo appears in the Australian coat of arms and holds a place of honor for many Australians. Other people simply are put off by the notion of native Australian cuisine, which conjures images of Aboriginal staples like grubs, peculiar foods that require effort to consume, rather than something to eat on a regular night out, like Italian.

Mark Olive is an Aboriginal cook who runs an Aboriginal food catering business located in a nondescript warehouse just a few blocks away from downtown Melbourne. He is widely known and has been featured in TV shows. When I visit Mark's catering business, I find him to be soft-spoken and charming, resembling a gentle bear. He once started an Aboriginal restaurant in Sydney, the Midden, but he says that the restaurant opened before the public was prepared to accept Aboriginal foods.

His current business offers an impressive range of native herbs and fruits: bush cucumber, desert lime, spicy desert raisins (kutjera), lemon myrtle leaf, marsdenia (bush banana), mountain pepperleaf and native pepperberry, muntrie berries, native basil and thyme, passionberry, quandong, rivermint, saltbush, sea parsley (also called sea celery), tanami apples, and the wattleseed that found its way into my meal. Ironically, the biggest consumers of Mark's herbs and spices are overseas buyers. Mark is dismayed by Australians' reluctance to recognize the bounty of native plants and animals and their inability to see popular "cute" animals as food.

"We have to get our own people in our own country to start utilizing more and more of these herbs and spices. We've got kangaroo, emu, and crocodile in this country that people tend to walk away from. I think that's because it is our coat of arms. For Aboriginal people, it was never their coat of arms. Just like sheep, pigs, everything else, it's a food source. Yes, they're cute, but I think lambs are cute, yet we eat them." Throughout much of recent Australian history, Aboriginals did not exist, in political life or civic society, and so their cuisine was also ignored. "If you weren't counted, you weren't part of the country," Mark says. "It wasn't until 1967 that we got birth certificates to say we're actually here. There have been big changes. Australia still has a long way to go, I think, in owning its Aboriginal history, being proud of it, for our immigrants to understand the history of this country. These sorts of things have to change."

Jon Belling, an Aboriginal man who works with Mission Australia, the group that runs Charcoal Lane, expresses similar frustrations. When I meet him at his office in downtown Melbourne, he looks placid and is exceedingly courteous, but as he begins to speak, something inside him uncorks.

"One of the nicest things we had here was wallaby burger that we did here for NAIDOC Day"—a reference to the National Aborigines and Islanders Day Observance Committee celebration of indigenous heritage in Australia. "We put it on the menu. I had some Aboriginal apprentices from one of our programs come and cook it. We had some chicken as well because '*WALLABY???*' Some people loved it; some people didn't even want to try it. They are cute, furry, cuddly little things that we throw food to, or look at." Like Mark, Jon laments Australians' inability to appreciate their native resources. "We have a treasure trove of food here in this country. We have companies coming here from overseas, coming from the States, looking at what we have here and taking it back. Australia has always been hesitant to look inward. Don't acknowledge the people—why would you acknowledge the product? Chefs from Spain and Germany are just crying for this stuff because they can see what it is."

Jon launches into a passionate speech on the virtues of Australian finger lime. As we leave his office, he tells me that he has just been diagnosed with type 2 diabetes—the result, he believes, of eating a Western diet. His doctor told him that Aboriginals are especially prone to an aggressive form of the disease. Aboriginal peoples were not exposed to Western diets and lifestyles until recently, and it is possible that their genes are less acclimated to aspects of Western diets such as high-glycemic foods (white flour, for example), which lead to greater increases in blood sugar, or lifestyles such as sedentary living. Australian populations with a greater mixture of European genes have lower rates of diabetes, which is consistent with the hypothesis of Aboriginals having greater genetic susceptibility.[13]

Australia is struggling with acceptance of native foods that could be more sustainably harvested and perhaps offer better nutritional value than imported animals and plants. What is the situation like in Canada or the United States? Do Canadians and Americans similarly disdain their native animals and plants? When I return to my home province of Ontario after almost two years abroad, I am eager to try native Canadian foods. The area where I grew up used to be the territory of First Nations peoples like the Algonquin, Iroquois, and Cree. In the north, the Cree hunted bear, deer, beaver, and waterfowl; to the east, the Algonquin and Iroquois supplemented game with corn, squash, beans, and wild rice. To the west, the Plains Indians followed herds of bison, and in the far west, salmon runs sustained the Pacific Coast Indians. In the far north, the Inuit hunted caribou, seal, whale, and fish; along the East Coast, the Micmac harvested shellfish, fish, and beaver. When I tell shopkeepers and butchers around Ottawa that I am looking for bear, beaver, and other game, however, they are baffled. One butcher in a trendy yuppie neighborhood just outside downtown Ottawa advises me to try a Chinese butcher shop in China-town, but the Chinese butchers are also unable to help.

The paucity of game around North America is due to a historical legacy of white settlers hunting many species almost to extinction. As a young man, Theodore Roosevelt went on a hunting trip to North Dakota, but the buffalo were nowhere to be found. An avid hunting enthusiast, Roosevelt founded the Boone and Crockett Club in 1887 to help conserve wildlife and promoted the use of science in wildlife management, which became known as the Roosevelt Doctrine.[14] The U.S. government eventually enacted laws that forbade the distribution of game for commercial purposes; the disappearances of the American bison, passenger pigeon, heath hen, and Carolina parakeet were alarming examples of what could happen to remaining species.[15] Canada followed suit, though Newfoundland and Nova Scotia permit the sale of game in restaurants, and Quebec is experimenting with looser restrictions. On its face, banning the trade in wildlife seems like sensible policy; without such an aggressive policy, the wilds of North America today would resemble Europe, virtually devoid of large game.

I drive to the outskirts of Ottawa on a cool spring morning to meet with a hunter, Kyle Worsley, who kindly donates some game meats for me to try, left over in his freezer from last year's fall hunt: bear, deer, moose. Head shaved bald, and gentle-spoken, Kyle comes from a family with a long tradition of hunting. He prefers hunting by bow and arrow. "Part of why I like the archery hunting versus gun hunting is you tend to see more wildlife. If you sneak as quietly as you can into the woods, you see a lot of different things walk by. Things that aren't necessarily in season that you are targeting. They just happen to wander by. I saw a wolverine once. I've seen wolves, coyotes, bears, just about every different creature that wanders through the Canadian wood."

Kyle, who runs a utility trailer business, heads out to the woods every fall to hunt deer for weeks at a time. Some hunters, Kyle notes, buy hunting tags but allow the prey to pass by, simply to enjoy the experience of hunting, rather than the killing.

So why not just leave the bow and arrow or gun at home and sit in the woods, I ask him.

"It's not the same," he replies. "There's an adrenaline rush when you see the animal walk up and you have the opportunity to shoot it. Even if you don't shoot it, the adrenaline rush is still there. If all you're doing is sitting in the woods, you're not as attuned, you're not listening for footsteps, you're not there with a purpose. When you're hunting, you're there with a purpose. You're looking for movement, you're listening for noises, you're trying to spot the game you're after."

Kyle understands his prey deeply; he discusses sex-ratio imbalances among deer due to misallocation of deer hunting permits with the keen observation of a naturalist. I wonder to myself: If someone had to be trusted with taking care of the woods and their wildlife, should it be a bureaucrat, a politician pandering to populist rhetoric about animal rights and gun control (Kyle seethes when he describes Liberal Party policies on gun control), or a hunter like Kyle, who surveys his domain in the woods week after week while fending off bugs? The wildlife in the woods would likely be in better shape if hunters had more voice in wildlife management.

Kyle supports the ban on wildlife trade, though. Roosevelt, instrumental in setting into motion the preservation of America's wildlife and wilderness, was himself a big-game enthusiast. No one could be more motivated to preserve the wildlife and wilderness than a person who derives great joy from them. Because of this wildlife-trade ban, however, enjoying the taste of native game is not easy. If you want to eat bear or beaver, as the Algonquin Indians did for thousands of years in the woods around eastern Ontario, you either have to buy a permit and shoot or trap the animal yourself, find a kind person who is willing to part with such meat (even bartering for wildlife is illegal—the regulations are uncompromising), or resign yourself to fantasizing about eating game meat while you're pushing a shopping cart down an aisle lined with factory-raised chicken, beef, and pork.

A trapper neighbor of my friend happens to have some frozen beaver leg sitting in a freezer from the previous fall. After soaking the leg in salt water overnight and slow-cooking it with wine and onions, I savor the dark, rich flavors of the meat. The meats from Kyle—bear sausage, moose, deer—also deluge me with a new palette of flavors. Once you start eating game and experience the variegated flavors, it is disappointing to go back to the blandness of supermarket meats.

The paradox behind my complex transaction in game meat and goodwill is that an animal that is enclosed by a fence, pumped with antibiotics, and fed grain that was doctored with insecticides, herbicides, and inorganic fertilizers is considered to be legal meat, but the animal on the other side of the fence, which by most measures is happier and healthier, has higher levels of omega-3 fatty acids, and is living in a more ecologically sustainable manner, cannot be bought or sold in a restaurant anywhere in most of North America.

It is very difficult to operate a restaurant serving native cuisine in North America, perhaps even more difficult than in Australia. In Ottawa, where museums enlighten visitors on everything from airplanes and trains to geology and the cultures of the First Nations, you might think a native-themed restaurant would be a hit. However, as I learned from Phoebe Blacksmith, buying elk or buffalo meat guarantees a stiff price markup that is passed on to the customer. I meet Phoebe on a rainy morning at a café. A Cree who is passionate about native Canadian cuisine, Phoebe (with her husband at the time) operated a restaurant called Sweetgrass in Ottawa's touristy downtown Byward Market area, serving native-themed dishes to curious customers. Over seven and a half years, it garnered favorable reviews, but the long hours took a toll. The couple divorced. Phoebe struggled for a year more on her own but arrived at work one day to find that she had been locked out—behind on rent. Phoebe was allowed to reenter the premises to claim perishable food. She packed it into her car, drove north, and crashed with relatives. She tried to open another restaurant to make use of the supplies left over from Sweetgrass, but the effort was futile. She declared bankruptcy; it took three long years to resolve the debts.

Phoebe still loves moose and goose meat; her mother gave birth to her while out in the bush during the spring goose hunt. Phoebe grew up eating everything the land had to offer, such as cranberries, chokeberries, strawberries, and crowberries. After Sweetgrass closed,

Phoebe went back to the land where she grew up, a reserve on Mistassini, the largest freshwater lake in Quebec. "Picking berries on the land healed me," she tells me. She got a call to teach native cooking at a college in northern Quebec, then went to the Cree community of Oujé-Bougoumou, population 725, to run a hotel-restaurant. Phoebe was appalled at the oily fries, burgers, and pizza that were staples on the menu and tried to get customers to eat buffalo meat. After the isolation and small-town gossip got to her, she moved back to Ottawa. Now she does catering, studies food hospitality and management at college, and dreams of opening a new food venture featuring native Canadian cuisine.

Hunters in North America are generally not in favor of the idea of allowing game to be sold. Recreational hunters fear that opening the markets to commercial hunting would mean more competition for game and hence less game available for them. Recreational hunters also worry that putting a dollar value on game meat would increase the temptation for poaching, which would further reduce game numbers. However, the ecological and nutritional benefits of a robust wildlife population are considerable; fare like deer, moose, beaver, bear, squirrel, and alligator could comprise an authentic Paleo diet of richly flavored meat, along with ecologically friendly sides like acorns, caterpillars, grasshoppers, and wild rice. But how can the benefits of wild foods be passed on to people in the United States and Canada who don't have the means, knowledge, or inclination to hunt or gather? Hunters in North America suggest that an attractive alternative is to move wild animals into domestic quarters; a second Agricultural Revolution, if you will, but this time done with more thought given to ecological and ethical consequences. At least, that's the hope, but the actual practice of moving wild animals from forests and oceans into confined quarters is fraught with challenges.

From an aerial photograph, you might think that the forests ringing Bearbrook Farm, on the outskirts of Ottawa, are maintained for timber or an adequate watershed, or perhaps for aesthetic reasons. It isn't until you drive down the long lane that bisects the farm and park your car at the end of the road that you finally spot the elk, skittish among the trees. Walter Henn, a tall, thick-boned fellow, manages Bearbrook Farm along with his wife Inge. While wind howls beyond the farmhouse doors, Walter tells me that as a consequence of the hardiness of the elk, bison, and deer that he rears, he doesn't have to medicate the animals with antibiotics; his animals never get sick.

"We concentrate on raising all of our animals as natural and humanely as possible. We don't use any chemicals on our farm. We do not use any chemicals for fertilizing. We only use manure for fertilizer. We do not use any chemicals for weed control. We clip all of the weeds. Most important of all, we let all of our animals run outside in their natural environment with the sun and wind and rain and everything." This philosophy extends to a desire for people to see his farm up close. Walter continues, "We invite all of our customers to come and visit our farm and see for themselves how the animals are being kept and being fed and enjoying the natural outdoor environment. Most farmers would not want to invite visitors because they raise their animals in closed environments like cages and locked-up barns as opposed to the natural way. They're also concerned about possibly spreading a disease if they have half a million chickens in a couple of chicken coops. We don't mind visitors at all."

I ask Walter if there are any special challenges in raising his animals.

"Buffalo and elk can be very temperamental, very challenging. You have to be careful not to enter the field without being on a tractor or staying on the outside of the fence because it's possible they could attack you. It's not normal for them to do it, but when they're under stress, when maybe they have a baby, they may charge you to protect a baby."

Because he has forsworn the use of artificial fertilizers, insecticides, or herbicides, the maintenance requirements are lower, but the meat yields are also lower than could otherwise be gotten from an industrial operation, which relegates Bearbrook to the status of a hobby farm; that's ideal for Walter and his wife, an elderly couple in retirement.

Walter's main impetus for farming elk and bison is that he wants to eat meat that is free of chemicals and naturally raised. At seventy-five years of age, he's not looking to scale up operations aggressively.

Bearbrook's grounds are visited by children and seniors who come to gawk at the elk, otherworldly bison, turkeys, chickens, and white-tailed deer (peacock are raised for ornamental purposes). Bearbrook Farm is employer, food production zone, recreation area, and eco-system anchor, all rolled into one. Walter and his wife give people a chance to eat meat that closely resembles the local wildlife.

Walter came to Canada from Germany after he refused to be drafted into the army. He lost his father and three uncles to World War II, and he didn't want to learn how to kill people. Walter and his father-in-law were pioneers in establishing dairy herds in eastern Ontario, he says. He and his wife tried their hand at the supermarket, hotel, and restaurant equipment business, traveling the world to set up his equipment, and at one point opened a bed-and-breakfast. Viewed in the light of these many ventures, Bearbrook Farm is just another extraordinary chapter in the couple's career.

"Some people call me and my wife workaholics. We need to do things, to have a challenge before us, to have a reason to get up in the morning, to be active. We're not like some of those brain-dead people who go golfing. We don't believe in that. It's a waste of time for society when you could contribute something good and nice to mankind and the next generation. We call it our hobby because we love doing it." If Walter had attempted to start Bearbrook Farm a few decades ago, he would have had to rely on word of mouth or advertisements in magazines and newspapers. Now the Internet is playing a new and important role, as orders for his game meats—in addition to the animals raised on the farm, Bearbrook also offers exotic meats like snake, crocodile, kangaroo, and camel—come in from many parts of the province. Technology is changing the face of commerce, and this seventy-five-year-old retiree is at the forefront of a new-yet-old way to raise food.

We are commonly advised these days to eat more fish, for the sake of omega-3 fatty acids that could lower our risks of coronary heart disease, allergic diseases, and depression, among other things. Americans eat more than twice as much salmon today as they did in 1990, but this increase has been accompanied by considerable controversy. In 1997, the United States went from being a net exporter to a net importer of salmon, despite opposition from American salmon farmers that resulted in tariffs on Norwegian and Chilean salmon. Rivers along the Atlantic coast once teemed with wild Atlantic salmon, but these fish have all but disappeared as an economic force, with major losses caused by damming of rivers, changes in water temperature, and other forms of habitat destruction. Although Alaska is the major producer of wild salmon in North America, nearly all of the increased numbers of salmon finding their way onto American dinner plates come from farmed salmon imported from Canada, Chile, and Norway.[16]

To learn more about salmon aquaculture, I book a seat on a train from Halifax to Moncton, a small city on Canada's east coast. The train pulls into downtown Moncton two hours later; low, drab houses cling like barnacles to a grid of widely spaced roads. My host is there to meet me, beaming. Dounia is a marine scientist who specializes in lobsters. She did some shopping at the supermarket next to the train station while waiting for me.

"Is salmon okay for tonight?" she asks.

In all likelihood, Dounia's salmon purchase originated from Cooke Aquaculture. Salmon aquaculture was first developed in Norway starting around 1970, then brought over to North America in 1978 after a Canadian scientist observed its potential. In 1984, New Brunswick had five fish farms. High prices for salmon drove the expansion of the industry, so that by 1996 the number of fish farms had swollen to seventy-seven. However, Chilean-farmed salmon began to enter the U.S. market, and disease and parasite epidemics ravaged farmed salmon stocks. In an attempt by the New Brunswick provincial authorities to clean up the waters of the bays that held the salmon pens, operators were required to own at least two sites, to allow one site to lie fallow while the other site held the salmon. Given the additional expenses that this entailed, the salmon aquaculture business was consolidated into the hands of just a few operators. Cooke has become by far the largest player in east-coast Atlantic salmon, evolving from a single New Brunswick farm with five thousand salmon to a multimillion-dollar, multinational enterprise, raising salmon, bream, and sea bass in Canada, the United States, Chile, Spain, and Scotland.[17]

Thierry Chopin, a professor from the University of New Brunswick who conducts research in cooperation with Cooke Aquaculture on making aquaculture more environmentally friendly, picks me up on a cheerful maritime blue morning and drives us to the gates of Cooke's hatchery. We step in and out of sterilization pools and scrub our hands with sanitizers repeatedly before entering. Salmon that are destined to be breeders swim in a spacious circular pool, with something of the leisurely atmosphere of a YMCA facility. Most of the salmon end up in outdoor, open-water pens for maturation—a highly controversial method, I have discovered.

We drive down to a small port and clamber into a boat with three Cooke employees. The boat motors out to a series of circular enclosures, a few hundred yards from shore. The nets of these enclosures hold thirty thousand to fifty thousand salmon per pen, depending on the size of the fish. Selectively bred to grow faster than wild Atlantic salmon, farmed salmon have been known to escape from their enclosures at certain facilities, through tears in the netting or accidental release into the surrounding waters during transfers. There are fears that escaped farm salmon could breed with local wild salmon, causing the gene pool to become weaker and pushing the wild stocks closer to extinction. In the outdoor pens, uneaten food and salmon feces drop onto the seafloor. One report estimated that the discharge of salmon feces into the Bay of Fundy from the aquaculture industry in 2005 was equivalent to the bowel movements of 93,450 people.[18]

A mat of white bacteria thus gathers beneath the pens, polluting the water with sulfides and causing oxygen levels to drop; little except hardy worms may live in these toxic environments. In addition, the fish exist in such crowded conditions that they are more easily infested with sea lice parasites, causing unsightly blemishes. Since blemished flesh is shunned by customers, fish farm operators are compelled to treat the sea lice outbreaks with pesticides, which can be poisonous for nearby animals such as lobsters. The sea lice may also transfer infectious salmon anemia, a disease that can be fatal to salmon and has wiped out huge stocks of farmed New Brunswick salmon in past years. When the province paid out compensation to fish farm operators like Cooke, there was a public outcry against the misuse of public funds. Opinions among fishermen and Native groups are complex: some of them decry the pollution and competition from salmon aquaculture operations, but others work in the industry itself and rely on aquaculture for steady incomes. The controversy over salmon aquaculture is most vociferous in North America, particularly the Pacific Northwest, which has the largest concentration of salmon pens; in Chile and Norway, governments are more lenient toward salmon aquaculture, and there is more space available for salmon farming operations, easing tensions and increasing profitability.

Although salmon aquaculture is new, aquaculture was known a thousand years ago to the Chinese, who raised carp in ponds, a practice that spread to Europe in the Middle Ages. Nowadays, basa and tra fish are raised in Southeast Asian ponds. Carp, basa, and tra are suitable fish to raise in ponds because they eat a broad variety of foods, including plant foods and human wastes, enabling recycling of valuable nutrients. Catfish have the same potential, and many are farmed in the States. All these fish, however, are challenging to export to Western markets because the fish have a muddy taste in their flesh, and the numerous small bones of carp make them difficult eating for people unaccustomed to the chore of picking out bones. On the other hand, once you develop a taste for carp and basa, they can be addictive. As I learned from living in China and Vietnam, Chinese people revere their carp, steamed or fried, bones be damned, and the Vietnamese simmer basa in soy sauce with ginger and garlic until it becomes pleasantly caramelized; the fat of the basa fish leaves a pleasant feel in the mouth alongside a dish of rice, never mind that these fish may have been fattened on human excrement in fish ponds.

Environmentalists would be far happier if salmon were raised in ponds that were inland instead of offshore, because the fish wastes and diseases would be contained more easily and escapees would be less of an issue. But raising a big fish like salmon in a pond is expensive: Salmon raised in recirculated water have a terrible taste, so rearing edible salmon requires either great quantities of freshwater or more extensive and expensive water treatment. Environmentalists counter by saying that open-water fish farms are polluting the oceans without paying, so it's only fair that aquaculture companies should bear these costs.

But it's not just the aquaculture companies who would absorb the costs—consumers would have to pay a premium for the privilege of eating salmon raised inland. The main reason that I sat down to a dinner of salmon with Dounia and her friends was that the salmon had been farmed and was therefore affordable. As a marine scientist working on lobsters, Dounia knew about the problems of farmed salmon, but she didn't have much choice. There was a fish market near her house, but it was only open when she was working. The same problem applies to sushi bars, which most commonly serve farmed salmon because it is easier to ship and keep fresh. Omega-3 fatty acids, which are easily damaged by heat and spoil quickly, are best obtained from fresh fish. Cooke Aquaculture prides itself on delivering fresh fish to consumers. The company's biggest asset is its proximity to the major consumer markets in eastern North America, cities like Toronto, Montreal, and New York City; the salmon that ends up on supermarket shelves arrives within forty-eight hours. For many chefs passionate about serving the freshest fish available, farmed salmon is the most popular option.

My tour continues to the Cooke fish-processing plant. The speed at which the salmon are processed is astonishing. The fish whiz along conveyer belts, and a well-groomed team pulls out fish parts when they face the wrong direction, jam up machines, or appear unsightly. After the heads are sliced off (the fish were killed by pneumatic gun after being pulled out of the pens), the bodies are sliced in half, the fins and bones are removed, the skin is descaled, and leftover bones are picked out by a platoon of workers. The workers (many of them are from the

Philippines and Romania and were hired on guest worker programs) look a little grim: The noise in the plant is deafening, the air chilly, and the gorgeous weather outside a fantasy for workers on twelve-hour shifts, but these are valuable jobs, and the premises are exactingly clean. There is hardly any of the fishy smell one might expect with a fishmonger or fish factory.

Later that evening, two Cooke reps, Chuck Brown, the communications manager, and Michael Szemerda, a vice president at the location, sit down with Thierry and me over dinner. The seared salmon is among the best that I have ever tasted, smooth and free of fishy tang. The flesh is a pleasing pink, due to a food-coloring carotenoid called canthaxanthin that is added to farmed salmon feed (and also chicken feed, to give an orange pigment to egg yolk and chicken fat). Wild salmon are additive-free, obtaining their carotenoids from krill. Chuck and Michael acknowledge that their business model is not perfect and that better environmental measures have to be instituted. Monterey Bay Aquarium and SeaChoice, a Canadian seafood program, issue three levels of recommendations for seafood: Green=Best Choice, Yellow=Some Concern, Red=Avoid. Atlantic salmon is labeled Red: Avoid. The decision disappoints Michael, who grumbles, "They try to paint everything with the same brush. Some of the reasons why they mark you as Red have absolutely nothing to do with us. Someone in Chile grows Atlantic salmon, which is not something that's native to Chile, but Atlantic salmon is native to the east coast of Canada! Everyone gets a Red."

Thierry runs a project with Cooke in an effort to mitigate environmental concerns. He grows seaweed near the salmon pens. The concept, known as IMTA (integrated multitrophic aquaculture), centers on the idea of putting together plants and aquatic animals that work symbiotically. Thierry's plan is that seaweed and mussels will absorb the fecal and food wastes from the fish pens, recycling the nutrients and also providing another marketable product. Thierry has his work cut out for him, however. Aside from industrial uses, such as providing carrageenans that are widely used as food thickeners and stabilizers, seaweed does not play a significant role in most Western

diets—or not yet, at least. The new craze for sushi is gradually introducing Westerners to Japanese and Korean use of seaweed for crispy
rice wrappers, sour and spicy salads, and heartening soups. Thierry
points out that IMTA is more than just salmon, seaweed, and mussels; in theory, there are an infinite variety of plants and animals that
could be usefully employed in conjunction with aquaculture, cleaning up the environment and providing food and other useful industrial products. Thierry believes IMTA could be done in closed-water
systems as well.

Environmental groups are pushing hard to make aquaculture
farms move their salmon operations inland. I drive out on a drizzly
afternoon to meet with Inka Milewski, science advisor at the Conservation Council of New Brunswick, at her farm. Thin and thoughtful,
she shows me pictures that she took of an abandoned fish farm: The
seafloor beneath was covered in a filthy gray mat of bacteria; bubbles
of sulphide gas streamed to the surface. Inka says inland fish farms
are better than open-water fish pens, but she would prefer that there
be no aquaculture at all. "We can't play God with nature," she says
emphatically. Indeed, some studies have observed that inland fish
farms have just as great an environmental impact as open-water
fish farms, or worse, due to the energy and water inputs necessary to
sustain the inland farms.

But where does this leave us? If we go with a system like SeaChoice
and carry around a card whenever we buy seafood, the criteria seem
overwhelming. For instance, wild Alaskan salmon is labeled as good,
but Atlantic salmon and farmed salmon anywhere are bad, except
Coho land-farmed salmon from the United States, which is permis-
sible, and wild salmon from the Pacific Coast, which is designated as
Yellow, Some Concern. With cod, the criteria are even more obscure:
Consumers are advised to avoid Atlantic cod (from Canada) and
Pacific cod (from Russia and Japan), but Pacific longline-caught cod
from Alaska is considered okay, while Pacific bottom-trawl cod from
the United States or British Columbia carries the Some Concern
warning. Whew! And those are just two items out of a list of thirty-
four seafood species.

However, some sort of action is imperative if we want future generations to have the same opportunities to eat wild fish. Fish catches worldwide peaked in the late 1980s and have since declined. Scientists have pointed out an analogy with forests, in which developed countries were able to increase forest cover due to populations shifting from the countryside to towns and increased environmental awareness; similarly, developed countries have established relatively good control over fisheries within their borders, but the long-term prospects of fisheries in Africa, Latin America, and much of Asia are bleak.[19]

Consumer choice is a powerful tool in this regard. Fears of PCB and mercury contamination put a dent in appetites for salmon until the industry learned how to remove the pollutants. In a globalized marketplace, we can achieve better omega-3/omega-6 balance without devastating large fish stocks by selecting smaller animals within the aquatic food chain (such as smaller, bonier fish and jellyfish), by consuming more sustainable land-based animals that have better omega-3/omega-6 balances (like insects and free-ranging chickens that forage on insects), and replacing vegetable oils that are rich in omega-6 fatty acids, such as corn oil, with animal fats.

We have considered the obstacles that people face in obtaining healthier meat and fish: the mental blocks that make game meat (and insects) unpalatable to many people, the laws preventing the sale of game in North America, the burdensome costs of raising wild animals like elk and salmon in confined spaces, the pollution that accompanies farmed salmon. A new frontier in agricultural science that aims to circumvent the risks of dwindling populations and disease is the widespread use of genetically modified organisms (GMOs) to produce more robust and plentiful stocks of animals and plants. The use of salmon genetically modified to grow much faster is still in the exploratory phase and is being closely watched by environmental groups. GMOs are hotly contested in many places in the world, particularly outside North America. Is this reaction just knee-jerk fear, or are our neighbors in this world justified in their apprehension?

One might think that the most prudent attitude is that we don't know what the long-term consequences are of growing and eating GMO corn, soybean, rice, potatoes, cucumbers, tomatoes, sweet peppers, peas, and canola. Not enough studies have been done because these high-tech plants, which are more resistant to weed killers and pests, were only introduced starting in 1996.[20] Unfortunately, as with nutritional research on soft drinks and milk, GMO studies conducted by scientists with connections to the industry tend to find no harmful effects, while scientists without such ties are more likely to observe adverse events.[21]

For instance, academic scientists in France discovered that rats fed with corn genetically modified to produce insecticides and withstand herbicides showed signs of toxicity in their kidneys and livers; in Italy, adverse genetic effects on embryos born from parental rats that ate herbicide-resistant genetically modified soybeans were observed; a team in Denmark found differences in weights of the small intestine, stomach, and pancreas in rats that ate rice genetically modified with an insecticidal gene spliced from kidney beans. Not exactly smoking guns, but given the history of industry interference in safety matters, such studies seem to argue for follow-up research and caution in releasing genetically modified foods to the public. Furthermore, weed resistance to Roundup, the herbicide developed by Monsanto, is spreading among American farms. China has seen increases in secondary pest infestations as a result of reliance on genetically modified cotton engineered to resist moths. Because the genetic modifications concocted by biotech scientists represent the most rapid instances of plant evolution and ecosystem alteration ever to have taken place on Earth, the sensible thing to do would be to conduct more tests on long-term safety and ban genetically modified foods in the interim, the approach adopted by Europe, Australia, and New Zealand.[22]

The United States, Canada, and other countries in the Americas have chosen a different route, the one heading to the hills of Profit. Genetically modified foods have risen to comprise a huge swath of American and Canadian agricultural production. After less than two decades, 93 percent of soy, 90 percent of corn, 95 percent of sugar-beet, 93 percent of rapeseed, and 30 percent of alfalfa crops are GMOs.[23] Since soy and corn enter the food system through myriad ways, such as through high-fructose corn syrup and animal feed, virtually everyone in North America consumes GM foods on a daily basis. The U.S. and Canadian governments authorize biotech companies to conduct their own health studies and their farmers to sell GM foods in supermarkets without informing consumers. Americans and Canadians have overwhelmingly stated in polls that they want GM ingredients to be labeled as such; a poll in 2014 by *The New York Times* found 92 percent of respondents supported GM labeling.[24] Health and agricultural regulatory bodies in both countries insist that GM foods are safe for consumers and the environment; therefore, there is no need to comply with public will or to foist the inconveniences (and sale losses) of GM labeling on food producers.

Who knows, the American and Canadian regulatory bodies may be right in the long run: GM foods may turn out to be benign or even beneficial to health and the environment, the magic bullet that solves the problem of world hunger, provides insecticide-free crops, and addresses nutritional deficiencies through genetic wizardry. However, in the face of united public opposition and the uncertainty concerning long-term biological and environmental effects, the paternalism of the U.S. Food and Drug Administration, Health Canada, and other regulatory bodies on the issue of GM labeling is a breathtaking affront. Connecticut and Maine have GM labeling regulations in place, but these are toothless: Both states require that neighboring states enact similar laws before any GM labeling happens.[25] Vermont is set to become the first state in North America to require mandatory GM food labeling, if it can successfully face down court challenges from Big Agriculture.[26]

I fly out to visit a longtime friend at his farm east of Iowa City. When I arrive in my rental car at night, the setting seems blissfully idyllic, a cozy house set amid sweeping vistas of corn. Next day, the landscape looks more sci-fi set than country pastoral, with choppers and planes swooping low and pumping out pesticides and fertilizer over thousands of acres of GM corn and soybeans. The phalanxes of

corn are tightly packed and stiffly uniform. Jon and his family corporation don't plant any corn for food on their holdings, almost three
thousand acres—part of a larger trend in which almost half of the U.S.
corn crop is reserved for ethanol production.[27]

"Some of the GM corn is so sturdy it can shred tractor tires," Jon
notes. He leads me into a cavernous shed housing massive, gleaming
combines, trucks, tractors; all told, perhaps a million dollars' worth
of equipment. There's a tree-lined river near the house where people
fish from little boats; the river would be lovely if the waters weren't
murky brown from nitrates and other chemical runoff. Jon refuses to
eat fish from the river, but he knows people who do. His two boys love
to play on the ATV, but when a plane swoops by the house, Jon and
his wife yell at the boys and usher them into the house.

"Don't like the kids playing out here when they're putting out
pesticides," he tells me.

Like everyone else, however, Jon's family corporation uses herbicides and pesticides to allow for easier harvesting and greater profitability. A farmer out here makes a comfortable living, as long as the
crop prices stay high enough, and profits are rolled back into equipment purchases. To increase their profits, farmers are constantly on
the lookout for more land, which would increase their scales of economy. When an elderly farmer appears ready to pass away, inquiries
about purchasing his land arrive like vultures catching the scent of
carrion.

In the evening, Jon and I set off for a jog. Jon is in fantastic shape,
while my right calf cramps up and I have to slow down. The gravel
roads inflict pain on my feet, poorly shielded in my five-toed minimal-
ist shoes, with a hole in one toe for good measure. In a way, my slow
pace is a good thing, because Jon and I haven't seen each other for
almost a decade and have a lot to catch up on. The air is clear of city
smog. The sun sets dramatically over the prairie landscape. We make
a right off the main road and run past cornfields. A mile or so down
the dirt road, we make a left and run past more cornfields. Then another left. More cornfields. I imagine bringing up a family out here,

in this expansive biotech paradise. It could be a healthy, peaceful life-style, the quiet nights lit with an expanse of stars.

However, the heavy use of chemicals on the crops is depressing. The long-term effects of such intense chemical exposure are largely unknown, but it seems an avoidable risk, if one is willing to accept lower profit margins. Jon tells me that although buyers of corn and soybeans pay higher prices for organic crops, the lower yields and extra work involved make organic crops unappealing for his family corporation.

Another health challenge of living in rural Iowa is the necessity of ve-hicles and the consequent epidemic of physical inactivity. Jon is in tip-top triathlete condition, but most Iowans don't have Jon's passion for exercise and have to drive long distances to go anywhere, as I found out from hanging out with Jon's family. Jon's family-run farming corporation is mechanized from top to bottom, and like most other rural Iowans, his relations struggle with obesity. However, scattered around the United States are places where farmers of various reli-gious and ideological persuasions resist the onslaught of Big Ag and practice old-fashioned labor-intensive agriculture on small farms. Among the best-known of these traditional farming groups are the Amish. At first glance, many people might find the Amish to be bi-zarre and impractical in their choices, but I wondered how life ap-peared from the inside, to people who lived a life that eschewed technology and most modern conveniences, and whether the long-term health prospects of these people were better than those of people in mainstream society.

Friends of mine in Des Moines take me to visit Jonathan Stutzman, about the same age as my longtime buddy but long-bearded and, on this particular Sunday afternoon, barefoot. Jonathan greets us on his porch. On a farm of 240 acres an hour's drive south of Des Moines, he and his family grow sweet corn, tomatoes, peaches, cucumbers, grapes, and cantaloupe, along with beef cattle and a few dairy cows, without the use of tractors. He travels by horse and buggy to sell his produce at a market. They have no refrigeration, so canning is important for preserving leftovers—nothing goes to waste. Two long-haired blond boys watch us curiously and quietly (the Amish speak a German dialect in their homes, and hence children generally do not learn English until they go to school), then bring over their pet rabbits to show the visitors. Girls carry buckets to and fro and bring out watermelon for the guests. I ask Jonathan what his family does for entertainment. Jonathan glances at me in astonishment. "We don't have time for entertainment. Sleep is our entertainment!" With eleven children to rear (two have already married) and no electricity, Jonathan and his family toil day to night. On this particular day, however, a languid Sunday, the children have guests over at the house, singing psalms.

Jonathan and I chat about the practice in some of the stricter Amish communities of not permitting rubber tires on vehicles. He recounts a tragic accident, in which a car struck and killed a man on a tractor, a few miles down from Jonathan's farm.

"The driver was using an iPod," he says.

I hear the anger seeping around the edges of his words. I think about the collision between new and old, one society oriented toward the maximization of pleasure and the minimization of inconvenience, the other focused on community, family, and faith. The roots of the Amish can be traced back to the sixteenth century, when a group of reformers in Switzerland decided that Martin Luther's proposals resulted in insufficient separation of the church from the demands of the state. In particular, these Swiss radicals believed that a true interpretation of the teachings of Christ argued for the renunciation of violence; they also maintained that children were too immature to choose religion, and therefore teenagers had to decide whether they would accept the teachings of Christ. Only upon this confession would the teenagers undergo baptism, rather than as infants, as commonly practiced among Protestants and Catholics. For their beliefs, the Anabaptists (i.e., rebaptized) were subjected to torture and/or killed, which led to the scattering of Anabaptist offshoot groups around Europe and eventually to the New World. The most conservative of these were the

Amish; the Mennonites are a well-known, less conservative branch of the Anabaptists.

In order to maintain community cohesiveness, conservative Amish groups today renounce or voluntarily limit their access to most forms of technology, including electricity and personal possession of telephones. Since Amish also generally ban ownership of gas-powered vehicles, they end up walking a great deal. A pedometer survey of an Old Order Amish group in the province of Ontario found that the men walked an average of eighteen thousand steps per day, while women walked fourteen thousand steps per day, both far outstripping the average of four thousand daily steps taken by Americans. With all the walking and the farm chores required to survive, there isn't much time for lazing around: The Amish men and women studied sat down about three hours per day. As a result, the prevalence of obesity is extremely rare in this group. None of the men in the survey were obese, and only 9 percent of the women, compared to an obesity rate of around 15 percent among the general Canadian population and 30 percent among the general U.S. population. Walking and lack of sitting must be the reason that these Amish were slim, because their diet was typical North American farm fare of meat, eggs, potatoes, bread, and vegetables, and rich in fat and sugar; moreover, Amish living in areas where people engage in much less farm work, such as Ohio, had obesity rates as high as or even *higher* than the average American population. The Amish may not just be physically healthier than people in mainstream North America; they also seem to do better in measures of mental health. While the closely knit nature of Amish communities may seem overly restrictive to outsiders, Amish women can draw upon the support of their communities and families, and thus suffer from far less depression than most American women.[28]

If the lifestyles of conservative Amish seem extreme to us, bear in mind that just one hundred years ago, this was the lifestyle of the majority of people living in North America. In the 1930s, only 10 percent of U.S. farms had electricity, while motorized tractors made their debut in 1900. From the point of view of our genes, the Amish are familiar, living in a manner more consistent with what our genes were made to accommodate. Modern mainstream lifestyles, based on sitting at a desk or in a car, interacting with many casual acquaintances and strangers but lacking in strong emotional support, and being shielded from the sun and the presence of common parasites, befuddle our genes and have spurred the rise of diseases like obesity, diabetes, depression, and allergic diseases. Our future well-being depends on whether we can recognize the monumental shifts that our lives have taken in just a hundred years and take action to restore our health.

# AFTERWORD

## *Rules to Eat and Live By*

Don't eat anything your great-grandmother wouldn't recognize as food.

—MICHAEL POLLAN, *Food Rules: An Eater's Manual*

My main goal in writing *100 Million Years of Food* was to explain what we should eat and how we should live by combining the latest in scientific studies on human nutrition and medicine with a dose of evolutionary biology and a review of how people past and present ate and lived. Over the course of this book I've examined a wealth of data, both scientific and anecdotal, about various lifestyles and diets. While many people are defined by their genes and the region where they live, there are certain universal truths about diet and health that apply to almost everyone.

## 1. KEEP MOVING

Although daily workouts and self-restraint in eating are commonly touted as the most effective means of avoiding food-related diseases like obesity and diabetes, neither scientific experiments nor an examination of human history supports these recommendations. Vigorous exercise makes people hungrier and often leads to physical injuries, while voluntarily cutting back on calories takes superhuman control and is probably an unnatural thing for humans to do, as our lean and fit hunter-gatherer relations had pretty hefty appetites. Instead, the most important thing you can do is to aim to walk like our ancestors and amble for at least two hours every day (around six to nine miles) or as much as feasible, or sit for a maximum of three hours a day, like the Old Order Amish mentioned in the previous chapter. Walking is free, requires no special equipment, and during summer daylight hours can also provide the benefit of sunlight/vitamin D exposure. Doing more moderate exercise for longer periods of time also cuts down on the health risks of sitting and watching TV.

To keep yourself motivated, find friends to walk with and get a pedometer or download a free app onto your smartphone to keep track of how many steps you take a day. In my case, walking for two hours results in around fourteen thousand steps, somewhat more than the current recommendation for ten thousand steps a day. You'll likely also experience a surge in spirits once you start walking regularly for two hours a day. A quick tip: Build up slowly to the two-hour goal. It is possible to injure yourself while walking, so better to take it easy for the first few months by taking shorter walks until you develop the necessary endurance. Carrying a little weight in your arms, like a water bottle or groceries, will give your upper body a suitably moderate workout as well. For people who lack the time to walk for two hours a day, adding as much walking, cycling, and other moderate exercise as possible and reducing time spent in front of a TV are sensible measures to take. New desk treadmills allow people in offices and libraries to walk while reading or typing.

## 2. DRINK ALCOHOL MODERATELY

Medical experts are generally conflicted over the merits of drinking alcohol because heavy drinking can harm the liver, increase the risk of metabolic syndrome, and increase the risk of violent death. On the other hand, when intake is moderate—two drinks a day for men, one drink a day for women—the health benefits of alcohol for mitigating heart disease and mortality in general are stronger than any known benefit from any other food item, including vegetables, fruits, and fish. That being said, the benefits of drinking alcohol accrue mainly to people over the age of forty living in developed countries, because infectious diseases rather than heart diseases tend to be the main killers in developing countries, and because for people under the age of forty, heart diseases are not an issue, whereas alcohol can exacerbate the risks that younger people face, such as accidents, homicides, and suicides.

## 3. EAT LESS MEAT AND DAIRY WHEN YOUNG

The current mainstream nutritional advice on meat is to eat sparing amounts; conversely, advocates of low-carb diets challenge the low-meat paradigm and assert that people should eat a lot of meat for better weight control and overall health, because starches are fattening and dangerous for heart health. Both sides are close to the truth. Younger people should eat less meat and dairy, because meat and dairy promote faster overall growth via hormones like IGF-1, which is a risk factor for certain types of cancers. On the other hand, for people over the age of sixty-five, eating more meat is likely a good thing, because the cancers that are promoted by meat take a long time to develop, whereas the real risk factors for an elderly person in the developed world stem from frailty and wasting, which may be mitigated by eating meat (dairy is more complicated, due to the high concentration of calcium). The common wisdom advises letting youth indulge in food and exercising restraint in later life, but this is exactly wrong; instead, we should advise younger people to eat meat and dairy sparingly, while people over the age of sixty-five should be encouraged to indulge in the pleasures of meat.

## 4. EAT TRADITIONAL CUISINE

While some food writers like Michael Pollan, Dr. Daphne Miller, and Sally Fallon Morell advocate eating some versions of traditional diets, most mainstream nutritionists are leery of traditional diets, which tend to be moderate in fat, cholesterol, and/or salt. I advocate traditional diets for three reasons: 1) In studies, traditional diets typically do at least as well as nutritionist-approved low-fat, low-salt diets in maintaining health. In part, this is because the functions of dietary fat, cholesterol, and salt throughout the body are numerous, while nutritionists have necessarily devoted their limited time and resources to narrow views on the harmful effects of these substances. 2) Traditional eaters didn't bother with scientific studies; they cooked and combined food in ways that maximized their health. The older the cuisine, the better: Five-hundred-year-old-cuisines are a good starting point, because at that point industrially processed foods had not yet made significant inroads into people's diets. 3) Traditional cuisines were moderate in fat, cholesterol, and/or salt and therefore tasted good; thus getting ourselves to stick with these diets is not difficult. The Mediterranean diet (olive oil, bread, nuts, goat cheese, fish, red wine, pasta, vegetables) is perhaps the most widely known and touted traditional cuisine these days, but many other traditional diets, from American southern and Mexican to Japanese, Okinawan (sweet potatoes, fish, vegetables, soybean), and Australian Aboriginal (kangaroo, crocodile, wild plants and fruits, tubers, honey), have been found to be superior to modern diets in mitigating chronic diseases like cancers and type 2 diabetes.

## 5. EAT WHAT YOUR ANCESTORS ATE

In societies where people lived on particular diets for hundreds or thousands of years, their bodies gradually became adapted to these diets, acquiring enzymes to process starches, in the case of Europeans and East Asians; to process seaweed, in the case of Japanese; and to process milk, in the case of northern Europeans, pastoralist African and Middle East groups, and northern Indians. High levels of calcium may be a risk factor for prostate cancer in populations that had little exposure to dairy. If your ancestors didn't consume much starch or dairy, neither should you. The take-home message: Eat what your ancestors ate.[1]

## 6. EAT SUSTAINABLY

Unfortunately, when we eat meat and fish cheaply, we do so by passing on the environmental costs of pollution and plant-cover degradation to future generations. The best way out of this mess is to eat more of the plants and animals that are adapted to our local environments and decrease our reliance on foreign, poorly adapted plants and animals. In many parts of the world, there is an abundance of plants and animals that people used to eat but later generations became squeamish about. In North America, acorns, deer, bear, moose, beaver, fish, waterfowl, and insects used to provide valuable sustenance, but European immigrants to the region rejected or forgot about these foods; kangaroo presents a similar dilemma for Australians descended from immigrants; insects are rejected in most of the developed world, and even tracts of the developing world. That's a shame, because wild plants and animals are generally better nutritional choices—for example, the ratio of omega-3 to omega-6 fatty acids is higher in wild foods—and more environmentally sustainable. Moreover, wild animals arguably live happier, more natural lives than their farm compatriots.

In the Americas, insects used to be a huge part of the ancestral diet because there weren't any big domesticated mammals around; they're still popular in much of the developing world. Insect protein is palatable. In Thailand, they can't import enough crickets to satisfy their demand. Pound for pound, insects consume fewer calories than mammal livestock because insects are cold-blooded. They also suck up less water and emit fewer greenhouse gases than livestock. Also, if you're worried about animal cruelty, the nervous system in an insect is far less developed than in a mammal.

## 7. GET AS MUCH SUN AS YOUR SKIN TYPE REQUIRES

Our ancestors were continually exposed to the sun. The most obvious manifestation of this is our body's dependency on skin exposure to sunlight to produce the correct amount of vitamin D. It's true that skin cancer is an opposing risk, so rather than getting burned on the weekend or hitting a tanning booth, the best thing to do is to spread out your exposure to the sun throughout the year and throughout the week, which allows people with tanning skin types to develop protective natural tanning. At the ends of the skin-type spectrum, people with light skin should be judicious in the intensity of sun exposure (think northern Europe), while people with dark skin should seek as much sun as practical. Solar radiation likely has an effect in reducing the risk of various types of cancers, such as breast cancer. Popping vitamin D pills or eating vitamin-D-rich food is not a great solution because scientists don't know how much vitamin D is required by the human body, or even if vitamin D is the main benefit from solar exposure; moreover, getting too much vitamin D may boost the risk of certain cancers, including prostate and colon cancer.

A final consideration with respect to sunlight is that in temperate regions of the world, cold weather may increase the risk of dying, independent of sunlight exposure—in other words, temperature is also an important factor in maintaining health.[2] Of course, in very warm locales, heat waves can be dangerous as well. In both cases, the elderly are most susceptible to the dangers of extreme temperatures. Thus if you are in your later years of life, want to optimize your health, and have the option of relocating, living somewhere where the temperature is congenial is a major health consideration.

## 8. GET SAFE GERM/PARASITE EXPOSURE

If you suffer from hay fever, food allergies, or other common immune system disorders, you can likely lay part of the blame on lack of sunlight (see the previous point) and the massive hygiene drive that started roughly one hundred years ago. Because our ancestors evolved with constant exposure to parasites like bacteria, viruses, and scores of tiny invertebrates, our immune systems are dependent upon parasitic exposure to calibrate properly, just as our teeth require hard foods, our feet require solid contact with ground, and our eyes require copious natural sunlight to develop properly. But parasites are no laughing matter, because many kinds of parasites can and will gladly finish us off; malaria, for example, kills 660,000 people worldwide each year, far outgunning the current deadly outbreak of Ebola. The challenge is to get enough exposure to parasites so that our immune systems develop properly, while avoiding mass epidemics due to unvaccinated children and adults. Studies of therapies employing parasites such as pig whipworms are currently undergoing FDA-scrutinized trials in the United States. There is a good case to be made that many antibiotic treatments are unnecessary and deplete the intestinal tract of helpful bacteria, so patients (and parents of children) facing antibiotic treatments should discuss with doctors which antibiotic treatments are necessary. Cesarean births may reduce the transmission of helpful bacteria from mothers to infants via vaginal secretions, so mothers should discuss with doctors the pros and cons of C-section deliveries and consider the use of swabs to apply vaginal smears to newborns.[3] Other options include spending more time in rural settings such as farms and traveling to developing countries.

## 9. COOK AT LOW HEAT

When a side of beef is roasted, a slab of salmon seared, a sliver of bacon fried, or a cube of tofu sautéed, a chemical process known as the Maillard reaction results in delicious browning of the cooked food (similar to caramelization). However, fatty or protein-rich foods cooked under high heat generate AGEs (advanced glycation end products). AGEs are also produced naturally in the body, but the concentration of circulating AGEs can be elevated through intake in industrialized diets. Like teenage pranksters, AGEs wreak havoc by binding to cell receptors, cross-linking and hence changing the shapes and functions of body proteins, and generally promoting oxidation damage and inflammation. Possible adverse health effects of AGEs include hardening of the arteries (atherosclerosis), anemia, Alzheimer's disease, cataracts, cirrhosis, bone brittleness, muscle stiffness, loss of grip strength, slower walking speed, kidney disease, type 1 and type 2 diabetes, and lowered life expectancy.[4]

Concentrations of AGEs can be altered enormously by different cooking techniques. Raw foods contain the fewest AGEs. Cooking using traditional, low-heat methods (boiling, steaming, stewing) produces slightly elevated levels of AGEs. High-temperature, dry methods of cooking (broiling, roasting, deep-frying, grilling) and food processing rack up the greatest yields of AGEs. Noxious AGEs are also highly prevalent in hamburgers, soft drinks, crackers, cookies, pretzels, doughnuts, pies, Parmesan cheese, pancakes, waffles, and other processed foods.[5]

## 10. REMEMBER: FAD DIETS DON'T WORK

Foods are one of the few things that we can easily alter in our lifestyles, and it's commonly believed that foods comprise the basis of our health—i.e., "You are what you eat." Not surprisingly, people gravitate toward various kinds of miracle diets and "superfoods" in the hopes of achieving a quick fix to health problems like obesity, diabetes, and cancers. However, eating more meat, or more dairy, or more fruits and vegetables, or more raw food, or less fat, or following any other dietary alteration has rarely provided relief from chronic diseases. There are two reasons for this lack of a quick dietary fix: 1) Our bodies are designed to thrive on a wide variety of foods, in the form of time-tested traditional diets. 2) The major factor underlying chronic disease is disruption in our physical lifestyles, particularly the absence of movement, so adjusting our diets to compensate for the lack of physical activity rarely achieves our desired goals. The final message: Eat good food, keep moving, and let your body take care of the rest.

# ENDNOTES

## INTRODUCTION: WHAT SHOULD WE EAT AND HOW SHOULD WE LIVE?

1. Khan et al., "Secular Trends in Growth and Nutritional Status of Vietnamese Adults in Rural Red River Delta after 30 Years (1976–2006)."
2. Fontana, "Long-Term Effects of Calorie or Protein Restriction on Serum IGF-1 and IGFBP-3 Concentration in Humans"; Gunnell et al., "Are Diet–Prostate Cancer Associations Mediated by the IGF Axis?"

## THE IRONY OF INSECTS

1. Eizirik, Murphy and O'Brien "Molecular Dating and Biogeography of the Early Placental Mammal Radiation"; Madsen et al., "Parallel Adaptive Radiations in Two Major Clades of Placental Mamals."
2. For recent views of the debate over the geographical origin of early primates, see, for example, Chaimanee et al., "Late Middle Eocene Primate from Myanmar and the Initial Anthropoid Colonization of Africa"; Perelman et al., "A Molecular Phylogeny of Living Primates"; and Springer et al., "Macroevolutionary Dynamics and Historical Biogeography of Primate Diversification Inferred from a Species Supermatrix."
3. Oonincx et al., "An Exploration on Greenhouse Gas and Ammonia Production by Insect Species Suitable for Animal or Human Consumption."

4. Paoletti et al., "Human Gastric Juice Contains Chitinase That Can Degrade Chitin."

5. Belluco et al., "Edible Insects in a Food Safety and Nutritional Perspective."

6. Raubenheimer and Rothman, "Nutritional Ecology of Entomophagy in Humans and Other Primates."

## THE GAMES FRUITS PLAY

1. Robbins et al., "Optimizing Protein Intake as a Foraging Strategy to Maximize Mass Gain in an Omnivore"; Rode and Robbins, "Why Bears Consume Mixed Diets During Fruit Abundance"; Levey and Rio, "It Takes Guts (and More) to Eat Fruit"; Izhaki and Safriel, "Why Are There So Few Exclusively Frugivorous Birds?"

2. Alinia, Hels, and Tetens, "The Potential Association Between Fruit Intake and Body Weight—A Review."

3. Haupt, "Ashton Kutcher's Fruitarian Diet."

4. Duboucher et al., "Pulmonary Lipogranulomatosis Due to Excessive Consumption of Apples."

5. Drouin, Godin, and Page, "The Genetics of Vitamin C Loss in Vertebrates."

6. Cui et al., "Recent Loss of Vitamin C Biosynthesis Ability in Bats"; Cui et al., "Progressive Pseudogenization."

7. Drouin, Godin, and Page, "The Genetics of Vitamin C Loss in Vertebrates."

8. Siegel, *Intoxication*; Hopkins, Bourdain, and Freeman, *Extreme Cuisine*; Whitten et al., *The Ecology of Sumatra*.

9. Levey et al., "Evolutionary Ecology of Secondary Compounds in Ripe Fruit."

10. Sadasivam and Thayumanayan, *Molecular Host Plant Resistance to Pests*.

11. Vissers et al., "Effect of Consumption of Phenols from Olives and Extra Virgin Olive Oil on LDL Oxidizability in Healthy Humans."

12. Bendini et al., "Phenolic Molecules in Virgin Olive Oils"; Hu, "The Mediterranean Diet and Mortality—Olive Oil and Beyond"; Kapellakis, Tsagarakis, and Crowther, "Olive Oil History, Production and By-Product Management"; Pérez-Jiménez et al., "The Influence of Olive Oil on Human Health"; Vossen, "Olive Oil."

13. Hu, "The Mediterranean Diet and Mortality—Olive Oil and Beyond"; Trichopoulou et al., "Adherence to a Mediterranean Diet and Survival in a Greek Population."

14. Steele, "Tannins and Partial Consumption of Acorns"; Altuğ, *Introduction to Toxicology and Food*; Kenward and Holm, "On the Replacement of the Red Squirrel in Britain: A Phytotoxic Explanation"; Serrano et al., "Tannins."

15. Heizer and Elsasser, *The Natural World of the California Indians*.

16. Bainbridge, "The Rise of Agriculture"; Clarke, *Edible and Useful Plants of California*; Bainbridge, "Use of Acorns for Food in California."

17. Diamond, *Guns, Germs, and Steel*.

18. Technically, each breadfruit is a collection of closely packed individual fruitlets that may each bear seed.

19. IICA, CARDI, and MINAG, *Seminar on Research and Development of Fruit Trees (Citrus Excluded).*; Motley, Zerega, and Cross, *Darwin's Harvest*; Wyatt, *All Your Gardening Questions Answered*.

20. Siler, "'Food of the Future' Has One Hitch"; D, *Breadfruit*.

21. Jones et al., "Isolation and Identification of Mosquito (*Aedes aegypti*) Biting Deterrent Fatty Acids from Male Inflorescences of Breadfruit (*Artocarpus altilis* [Parkinson] Fosberg)."

22. Taubes, *Good Calories, Bad Calories*.

23. Mensink et al., "Effects of Dietary Fatty Acids and Carbohydrates on the Ratio of Serum Total to HDL Cholesterol and on Serum Lipids and Apolipoproteins."

24. Stanhope and Prior, "The Tokelau Island Migrant Study."

25. Ostbye et al., "Type 2 (Non-Insulin-Dependent) Diabetes Mellitus, Migration and Westernisation."

26. Siemens et al., "Spider Toxins Activate the Capsaicin Receptor to Produce Inflammatory Pain."

27. Birds are unfazed by capsaicin. This could mean that chili plants use birds as a unique means of seed dispersal, while avoiding mammals like rodents that may digest the seeds and extinguish the plant's reproductive prospects.

28. Rozin and Schiller, "The Nature and Acquisition of a Preference for Chili Pepper by Humans."; Sherman and Billing, "Darwinian Gastronomy"; Billing and Sherman, "Antimicrobial Functions of Spices."

29. Rozin and Schiller, "The Nature and Acquisition of a Preference for Chili Pepper by Humans."

30. Solomon, "The Opponent-Process Theory of Acquired Motivation."

31. Yoshioka et al., "Effects of Red-Pepper Diet on the Energy Metabolism in Men."

32. Ludy, Moore, and Mattes, "The Effects of Capsaicin and Capsiate on Energy Balance"; Singletary, "Red Pepper."

33. "What Is CH-19 Sweet Pepper?"

34. Johnson et al., "The Planetary Biology of Ascorbate and Uric Acid and Their Relationship with the Epidemic of Obesity and Cardiovascular Disease."

35. Abdelgadir, Wahbi, and Idris, "Some Blood and Plasma Constituents of the Camel."

36. Marcus, *Kluge*.

37. Casas-Agustench, Salas-Huetos, and Salas-Salvadó, "Mediterranean Nuts."

38. Choi, Gao, and Curhan, "Vitamin C Intake and the Risk of Gout in Men."

39. Sutin et al., "Impulsivity Is Associated with Uric Acid."

40. Singer and Wallace, "The Allopurinol Hypersensitivity Syndrome"; Becker et al., "Febuxostat Compared with Allopurinol in Patients with Hyperuricemia and Gout."

41. Kratzer et al., "Evolutionary History and Metabolic Insights of Ancient Mammalian Uricases."

42. Hawkes, *The Labrador Eskimo*; Smith, *Inujjuamiut Foraging Strategies*.

## THE TEMPTATION OF MEAT

1. "Papua New Guinea."

2. Buettner, *The Blue Zones.*

3. Fryxell and Sinclair, "Causes and Consequences of Migration by Large Herbivores."

4. Hofreiter et al., "Vertebrate DNA in Fecal Samples from Bonobos and Gorillas"; Surbeck and Hohmann, "Primate Hunting by Bonobos at LuiKotale, Salonga National Park."

5. Hardus et al., "Behavioral, Ecological, and Evolutionary Aspects of Meat-Eating by Sumatran Orangutans (*Pongo abelii*)"; Hofreiter et al., "Vertebrate DNA in Fecal Samples from Bonobos and Gorillas."

6. Semaw et al., "2.6-Million-Year-Old Stone Tools and Associated Bones from OGS-6 and OGS-7, Gona, Afar, Ethiopia."

7. Hoberg, "Phylogeny of Taenia."

8. Trinkel, "Prey Selection and Prey Preferences of Spotted Hyenas *Crocuta crocuta* in the Etosha National Park, Namibia."

9. Liebenberg, "Persistence Hunting by Modern Hunter-Gatherers"; Bramble and Lieberman, "Endurance Running and the Evolution of *Homo*"; Cunningham et al., "The Influence of Foot Posture on the Cost of Transport in Humans"; Carrier et al., "The Energetic Paradox of Human Running and Hominid Evolution [and Comments and Reply]."

10. Wrangham, "Evolution of Coalitionary Killing."

11. Falk et al., "Early Hominid Brain Evolution"; Anton, "Natural History of *Homo erectus*."

12. Roebroeks and Villa, "On the Earliest Evidence for Habitual Use of Fire in Europe."

13. Lepre et al., "An Earlier Origin for the Acheulian."

14. Mithen, "'Whatever Turns You On': A Response to Anna Machin,'Why Handaxes Just Aren't That Sexy.'"

15. Gowlett, "Special Issue."

16. A similar hypothesis is that the hand-axes were given as valuable gifts, to cement ties with important allies, but whether this constituted the principal function of hand-axes for more than a million years over a large geographical region also stretches the imagination.

17. Davidson, "Australian Throwing-Sticks, Throwing-Clubs, and Boomerangs"; Isaac, "Throwing and Human Evolution."

18. Whittaker and McCall, "Handaxe-Hurling Hominids."

19. The argument could be made that Stefansson and Andersen were both of Nordic ancestry, and therefore preadapted to meat-heavy diets through genes or childhood exposure.

20. Falchi et al., "Low Copy Number of the Salivary Amylase Gene Predisposes to Obesity."

21. Hopkins, "Effects of Dietary Cholesterol on Serum Cholesterol."

22. Davenport, *Aphrodisiacs and Anti-Aphrodisiacs*; Alcock, *Food in the Ancient World*; Delany, "Constantinus Africanus' *De Coitu.*"

23. Cheney, "The Oyster in Dutch Genre Paintings."

24. Delany, "Constantinus Africanus' 'De Coitu.'"

25. John Smith, "A Rhapsody upon a Lobster," in King, *Lobster.*

26. Barona and Fernandez, "Dietary Cholesterol Affects Plasma Lipid Levels, the Intravascular Processing of Lipoproteins and Reverse Cholesterol Transport Without Increasing the Risk for Heart Disease."

27. Casas-Agustench, Salas-Huetos, and Salas-Salvadó, "Mediterranean Nuts"; Aldemir et al., "Pistachio Diet Improves Erectile Function Parameters and Serum Lipid Profiles in Patients with Erectile Dysfunction."

28. De Graaf, Brouwers, and Diemont, "Is Decreased Libido Associated with the Use of HMG-CoA-Reductase Inhibitors?"; Schooling et al., "The Effect of Statins on Testosterone in Men and Women, a Systematic Review and Meta-Analysis of Randomized Controlled Trials."

29. Zhang, "Epidemiological Link Between Low Cholesterol and Suicidality."

30. Tamakoshi, Yatsuya, and Tamakoshi, "Early Age at Menarche Associated with Increased All-Cause Mortality"; Rogers et al., "Diet throughout Childhood and Age at Menarche in a Contemporary Cohort of British Girls."

31. Abbasi et al., "Experimental Zinc Deficiency in Man: Effect on Testicular Function"; Kynaston et al., "Changes in Seminal Quality Following Oral Zinc Therapy."

32. Elgar and Crespi, *Cannibalism*; Saladie et al., "Intergroup Cannibalism in the European Early Pleistocene."

33. Liberski et al., "Kuru."

34. Mead et al., "Balancing Selection at the Prion Protein Gene Consistent with Prehistoric Kurulike Epidemics."

35. Diamond, "Archaeology."

## THE PARADOX OF FISH

1. "Japan Bluefin Tuna Fetches Record $1.7m."

2. Simoons, "Fish as Forbidden Food"; Dobney and Ervynck, "To Fish or Not to Fish?"; Malainey, Przybylski, and Sherriff, "One Person's Food"; Simoons, *Eat Not This Flesh*; Buxton, "Fish-Eating in Medieval England"; Diamond, *Collapse*; Woolgar, "Food and the Middle Ages"; Pálsson, *Coastal Economies, Cultural Accounts*; Henrich and Henrich, "The Evolution of Cultural Adaptations."

3. Simoons, "Fish as Forbidden Food"; Simoons, "Rejection of Fish as Human Food in Africa."

4. Akazawa et al., "The Management of Possible Fishbone Ingestion"; Kodama and Hokama, "Variations in Symptomatology of Ciguatera Poisoning"; Lehane and Lewis, "Ciguatera"; Begossi, Hanazaki, and Ramos, "Food Chain and the Reasons for Fish Food Taboos Among Amazonian and Atlantic Forest Fishers (Brazil)."

5. Allport, *The Queen of Fats*; Usui et al., "Eicosapentaenoic Acid Plays a Role in Stabilizing Dynamic Membrane Structure in the Deep-Sea Piezophile *Shewanella violacea*"; Balny, Masson, and Heremans, *Frontiers in High Pressure Biochemistry and Biophysics*; Bell, Henderson, and Sargent, "The Role of Polyunsaturated Fatty Acids in Fish."

6. The two forms of omega-3 that are useful to humans are EPA (eicosapentaenoic acid) and DHA (docosahexaenoic acid). EPA and DHA are also present in the flesh, organs, eggs, and milk of animals that browse on natural diets, including grass, seaweed (EPA only), and insects. Humans can also synthesize limited amounts of both EPA and DHA from ALA (alpha-linolenic acid), which is found in the chloroplasts of wild plants. Flaxseed oil, for instance, is a rich source of ALA. Omega-6 fatty acids also come in two main forms, linoleic acid (LA) and arachidonic acid (AA). LA is found in the seeds of most plants (with the exceptions of coconut, cocoa, and palm), and AA is found in meat and other animal products. Humans are able to synthesize AA (the meat-based omega-6) from LA (plant-based omega-6). Simopoulos, "The Importance of the Omega-6/Omega-3 Fatty Acid Ratio in Cardiovascular Disease and Other Chronic Diseases."

7. Simopoulos, "The Importance of the Omega-6/Omega-3 Fatty Acid Ratio in Cardiovascular Disease and Other Chronic Diseases"; Calder, "The Role of

Marine Omega-3 (n-3) Fatty Acids in Inflammatory Processes, Atherosclerosis and Plaque Stability."

8. Simopoulos, "The Importance of the Omega-6/Omega-3 Fatty Acid Ratio in Cardiovascular Disease and Other Chronic Diseases"; Eaton et al., "Dietary Intake of Long-Chain Polyunsaturated Fatty Acids During the Paleolithic"; Meyer et al., "Dietary Intakes and Food Sources of Omega-6 and Omega-3 Polyunsaturated Fatty Acids"; Sioen et al., "Dietary Intakes and Food Sources of Fatty Acids for Belgian Women, Focused on n-6 and n-3 Polyunsaturated Fatty Acids"; Sugano and Hirahara, "Polyunsaturated Fatty Acids in the Food Chain in Japan"; Pella et al., "Effects of an Indo-Mediterranean Diet on the Omega-6/Omega-3 Ratio in Patients at High Risk of Coronary Artery Disease"; Blasbalg et al., "Changes in Consumption of Omega-3 and Omega-6 Fatty Acids in the United States During the 20th Century"; Taubes, *Good Calories, Bad Calories.*

9. MacLean et al., "Effects of Omega-3 Fatty Acids on Cancer Risk"; Saynor, Verel, and Gillott, "The Long-Term Effect of Dietary Supplementation with Fish Lipid Concentrate on Serum Lipids, Bleeding Time, Platelets and Angina."

10. The waxy-leaf nightshade plant (*Solanum glaucophyllum*) produces very high levels of vitamin D, likely as protection against animal predation; animals that browse heavily on this plant suffer from hypercalcium, leading to calcification of tissues and possibly death.

11. Björn, "Vitamin D"; Lazenby and McCormack, "Salmon and Malnutrition on the Northwest Coast"; Maji, "Vitamin D Toxicity."

12. Simoons, *Eat Not This Flesh.*

13. Richerson and Boyd, "Built for Speed, Not for Comfort."

14. Plutarch, *Isis and Osiris.*

15. Simoons, *Eat Not This Flesh.*

16. Cerulli, *Peoples of South-West Ethiopia and Its Borderland.*

17. Saisithi, "Traditional Fermented Fish."

18. Curtis, "Umami and the Foods of Classical Antiquity"; Saisithi, "Traditional Fermented Fish."

19. Martial, *Epigrams.*

20. In Vietnamese: *Tuong Ban cham voi tai de / An vao mot mieng bung bung nhu de / Em oi, o lai dung ve / Ngay mai ta lai Tuong Ban tai de.*

21. Nakamura et al., "A Japanese Diet and 19-Year Mortality"; Goldbohm et al., "Dairy Consumption and 10-Y Total and Cardiovascular Mortality."

22. Kurihara, "Glutamate."

23. Shimada et al., "Headache and Mechanical Sensitization of Human Pericranial Muscles After Repeated Intake of Monosodium Glutamate (MSG)."

24. He et al., "Consumption of Monosodium Glutamate in Relation to Incidence of Overweight in Chinese Adults"; Insawang et al., "Monosodium Glutamate (MSG) Intake Is Associated with the Prevalence of Metabolic Syndrome in a Rural Thai Population."

25. Samuels, "The Toxicity/Safety of Processed Free Glutamic Acid (MSG)."

26. Mosby, "'That Won-Ton Soup Headache'"; Walker and Lupien, "The Safety Evaluation of Monosodium Glutamate."

27. Shi et al., "Adaptive Diversification of Bitter Taste Receptor Genes in Mammalian Evolution"; Huang et al., "The Cells and Logic for Mammalian Sour Taste Detection."

## THE EMPIRE OF STARCHES

1. Blount, *Soupsongs/Webster's Ark.*

2. Mintz and Schlettwein-Gsell, "Food Patterns in Agrarian Societies."

3. Feynman and Ruzmaikin, "Climate Stability and the Development of Agricultural Societies."

4. Price and Bar-Yosef, "The Origins of Agriculture"; Cohen, "Introduction."

5. Cardillo and Lister, "Death in the Slow Lane"; Roberts et al., "New Ages for the Last Australian Megafauna"; Holdaway and Jacomb, "Rapid Extinction of the Moas (Aves: Dinornithiformes)"; Roberts and Jacobs, "The Lost Giants of Tasmania"; Diamond, "Palaeontology"; Norton et al., "The Nature of Megafaunal Extinctions During the MIS 3–2 Transition in Japan"; Anderson et al., "Faunal Extinction and Human Habitation in New Caledonia."

6. Munro, "Epipaleolithic Subsistence Intensification in the Southern Levant."

7. Bar-Yosef, "Climatic Fluctuations and Early Farming in West and East Asia."

8. Larsen, "The Agricultural Revolution as Environmental Catastrophe."

9. Murgatroyd, *Dig 3ft NW*; Murgatroyd, *The Dig Tree*; Gregory, *Australia's Great Explorers*; Robson, *Great Australian Speeches*; Clarke, *Aboriginal Plant Collectors*; French, *The Camel Who Crossed Australia.*

10. Arditti and Rodriguez, "Dieffenbachia."

11. Roberts, *Margaret Roberts' A–Z of Herbs*; Kowalchik and Hylton, *Rodale's Illustrated Encyclopedia of Herbs*; Pohanish, *Sittig's Handbook of Toxic and Hazardous Chemicals and Carcinogens*; Gaillard and Pepin, "Poisoning by Plant Material"; Emsley, *Molecules of Murder.*

12. Turkington and Mitchell, *The Encyclopedia of Poisons and Antidotes*; Gaillard and Pepin, "Poisoning by Plant Material."

13. Turkington and Mitchell, *The Encyclopedia of Poisons and Antidotes*; Gaillard and Pepin, "Poisoning by Plant Material"; Barceloux, *Medical Toxicology of Natural Substances*; Eppinger, *Field Guide to Wild Flowers of Britain and Europe*; Gibbons,

Haynes, and Thomas, *Poisonous Plants and Ven Animals*; Bryson, *Comprehensive Reviews in Toxicology*; Nellis, *Poisonous Plants and Animals of Florida and the Caribbean*; Vizgirdas and Rey-Vizgirdas, *Wild Plants of the Sierra Nevada*; Lewis, *Lewis' Dictionary of Toxicology*; Kurian and Sankar, *Medicinal Plants*; Roberts, *Margaret Roberts' A–Z of Herbs*; Kowalchik and Hylton, *Rodale's Illustrated Encyclopedia of Herbs*; Emsley, *Molecules of Murder*; Greim and Snyder, *Toxicology and Risk Assessment*; Karmakar, *Forensic Medicine and Toxicology*; Iwu, *Handbook of African Medicinal Plants*; Panda, *Herbs Cultivation and Medicinal Uses*; Schmelzer and Gurib-Fakim, *Medicinal Plants 1*; Meuninck, *Medicinal Plants of North America*; Tilford, *Edible and Medicinal Plants of the West*; Fuller and McClintock, *Poisonous Plants of California*.

14. Lawley, Curtis, and Davis, *The Food Safety Hazard Guidebook*; Jha, "Man Dies After Drinking Lauki Juice."
15. Also known as khesari dal. The toxic effects of grass pea are due to accumulation of the toxic amino acid ODAP.
16. Bruyn and Poser, *The History of Tropical Neurology*; Rutter and Percy, "The Pulse That Maims."
17. Krakauer, "How Chris McCandless Died."
18. McMillan and Thompson, "An Outbreak of Suspected Solanine Poisoning in Schoolboys: Examination of Criteria of Solanine Poisoning"; "Solanine Poisoning."
19. Seigler, *Plant Secondary Metabolism*.
20. Fuller and McClintock, *Poisonous Plants of California Natural History Guides*; Deshpande, *Handbook of Food Toxicology*; Williamson et al., *Venomous and Poisonous Marine Animals*; Fenwick and Oakenfull, "Saponin Content of Food Plants and Some Prepared Foods."
21. Rea, Thompson, and Jenkins, "Lectins in Foods and Their Relation to Starch Digestibility."
22. Walters, *Plant Defense*; Arnoldi, *Functional Foods, Cardiovascular Disease, and Diabetes*; Deshpande, *Handbook of Food Toxicology*; Ayyagari, Narasinga Rao, and Roy, "Lectins, Trypsin Inhibitors, BOAA and Tannins in Legumes and Cereals and the Effects of Processing"; Riemann and Cliver, *Foodborne Infections and Intoxications*; Bewley, Black, and Halmer, *The Encyclopedia of Seeds*.
23. Vasconcelos et al., "Detoxification of Cassava During Gari Preparation"; Tylleskär et al., "Cassava Cyanogens and Konzo, an Upper Motoneuron Disease Found in Africa"; Haque and Bradbury, "Total Cyanide Determination of Plants and Foods Using the Picrate and Acid Hydrolysis Methods"; Satya et al., "Bamboo Shoot Processing."
24. Packard, *Processed Foods and the Consumer*.

25. Lott et al., "Phytic Acid and Phosphorus in Crop Seeds and Fruits"; Libert and Franceschi, "Oxalate in Crop Plants"; Siener et al., "Oxalate Content of Cereals and Cereal Products"; Porth, *Essentials of Pathophysiology*; Duhan et al., "Phytic Acid Content of Chickpea (*Cicer arietinum*) and Black Gram (*Vigna mungo*)"; Bishnoi, Khetarpaul, and Yadav, "Effect of Domestic Processing and Cooking Methods on Phytic Acid and Polyphenol Contents of Pea Cultivars (*Pisum sativum*)"; Reddy and Pierson, "Reduction in Antinutritional and Toxic Components in Plant Foods by Fermentation"; Savage et al., "Effect of Cooking on the Soluble and Insoluble Oxalate Content of Some New Zealand Foods."

26. Zohary, Hopf, and Weiss, *Domestication of Plants in the Old World*.

27. Bower, Sharrett, and Plogsted, *Celiac Disease*; Smith, *Celiac Disease*; Zhernakova et al., "Evolutionary and Functional Analysis of Celiac Risk Loci Reveals SH2B3 as a Protective Factor Against Bacterial Infection."

28. Bower, Sharrett, and Plogsted, *Celiac Disease*; Smith, *Celiac Disease*; Zhernakova et al., "Evolutionary and Functional Analysis of Celiac Risk Loci Reveals SH2B3 as a Protective Factor Against Bacterial Infection"; Sapone et al., "Spectrum of Gluten-Related Disorders."

29. Zhernakova et al., "Evolutionary and Functional Analysis of Celiac Risk Loci Reveals SH2B3 as a Protective Factor Against Bacterial Infection"; Haboubi, "Coeliac Disease: From A–Z"; "Being Gluten-Free 'Is Determined by Evolution', Says Gastroenterologist."

30. Velasquez-Manoff, "What Really Causes Celiac Disease?"

31. Decker et al., "Cesarean Delivery Is Associated with Celiac Disease but Not Inflammatory Bowel Disease in Children."

32. Sapone et al., "Spectrum of Gluten-Related Disorders"; Catassi et al., "Non-Celiac Gluten Sensitivity."

33. Peters et al., "Potential Benefits and Hazards of Physical Activity and Exercise on the Gastrointestinal Tract"; Johannesson et al., "Physical Activity Improves Symptoms in Irritable Bowel Syndrome"; de Oliveira and Burini, "The Impact of Physical Exercise on the Gastrointestinal Tract"; Gibson and Shepherd, "Food Choice as a Key Management Strategy for Functional Gastrointestinal Symptoms."

## ELIXIRS

1. Freedman et al., "Association of Coffee Drinking with Total and Cause-Specific Mortality."

2. Catling et al., "A Systematic Review of Analytical Observational Studies Investigating the Association Between Cardiovascular Disease and Drinking

Water Hardness"; Monarca et al., "Drinking Water Hardness and Cardiovascular Disease."

3. McGovern et al., "Fermented Beverages of Pre- and Proto-Historic China."
4. Levey, "The Evolutionary Ecology of Ethanol Production and Alcoholism."
5. Piškur et al., "How Did *Saccharomyces* Evolve to Become a Good Brewer?"
6. Kinde et al., "Strong Circumstantial Evidence for Ethanol Toxicosis in Cedar Waxwings (*Bombycilla cedrorum*)"; Dennis, "If You Drink, Don't Fly."
7. Marmot, "Alcohol and Coronary Heart Disease"; Bovet and Paccaud, "Commentary"; Marmot, "Commentary"; Rimm et al., "Moderate Alcohol Intake and Lower Risk of Coronary Heart Disease"; Stec et al., "Association of Fibrinogen with Cardiovascular Risk Factors and Cardiovascular Disease in the Framingham Offspring Population."
8. Bovet and Paccaud, "Commentary"; Marmot, "Commentary"; Bremer, Mietus-Snyder, and Lustig, "Toward a Unifying Hypothesis of Metabolic Syndrome."
9. Peng et al., "The ADH1B Arg47His Polymorphism in East Asian Populations and Expansion of Rice Domestication in History."
10. Prentice, "Diet, Nutrition and the Prevention of Osteoporosis"; Grivas et al., "Association Between Adolescent Idiopathic Scoliosis Prevalence and Age at Menarche in Different Geographic Latitudes."
11. Jouan et al., "Hormones in Bovine Milk and Milk Products: A Survey."
12. Leonardi et al., "The Evolution of Lactase Persistence in Europe. A Synthesis of Archaeological and Genetic Evidence."
13. Malacarne, "Protein and Fat Composition of Mare's Milk."
14. Salimei and Fantuz, "Equid Milk for Human Consumption"; Malacarne, "Protein and Fat Composition of Mare's Milk"; Faye, "The Sustainability Challenge to the Dairy Sector—The Growing Importance of Non-Cattle Milk Production Worldwide."
15. Fessler and Haley, "Guarding the Perimeter"; Gade, "Llamas and Alpacas"; Gade, *Nature and Culture in the Andes.*
16. Curtis, Aunger, and Rabie, "Evidence That Disgust Evolved to Protect from Risk of Disease."
17. Ottaviani, Camera, and Picardo, "Lipid Mediators in Acne."
18. Hegsted, "Fractures, Calcium, and the Modern Diet"; Prentice, "Diet, Nutrition and the Prevention of Osteoporosis."
19. Koh et al., "Gender-Specific Associations Between Soy and Risk of Hip Fracture in the Singapore Chinese Health Study"; Nimptsch et al., "Dietary Vitamin K Intake in Relation to Cancer Incidence and Mortality"; Chow, "Dietary Intake of Menaquinones and Risk of Cancer Incidence and Mortality."

20. Rowland et al., "Calcium Intake and Prostate Cancer Among African Americans"; Kretchmer et al., "Intestinal Absorption of Lactose in Nigerian Ethnic Groups."

21. Sellers, Sharma, and Rodd, "Adaptation of Inuit Children to a Low-Calcium Diet."

22. Fediuk et al., "Vitamin C in Inuit Traditional Food and Women's Diets"; Njoku, Ayuk, and Okoye, "Temperature Effects on Vitamin C Content in Citrus Fruits"; Jacobs, *The Pastoral Masai of Kenya*; Dickson, *The Arab of the Desert*; Burckhardt, *Notes on the Bedouins and Wahábys*; Leshem et al., "Enhanced Salt Appetite, Diet and Drinking in Traditional Bedouin Women in the Negev"; Wagh et al., "Lactase Persistence and Lipid Pathway Selection in the Maasai."

23. Lactase persistence has been observed at frequencies of more than 80 percent among Tutsis, Beja, Tuareg, and Bedouins; 23 percent to 76 percent of Jordanians; and 22 percent to 86 percent of people in Saudi Arabia. Leonardi et al., "The Evolution of Lactase Persistence in Europe. A Synthesis of Archaeological and Genetic Evidence."; Heyer et al., "Lactase Persistence in Central Asia."

24. Wiley, *Re-Imagining Milk*; DuPuis, *Nature's Perfect Food*; Elliott, "Canada's Great Butter Caper: On Law, Fakes and the Biography of Margarine."

## A TRUCE AMONG THIEVES

1. Barnes, *Diseases and Human Evolution.*

2. "Preservation of Health in the Japanese Navy and Army."

3. Carpenter, *Beriberi, White Rice, and Vitamin B.*

4. Rajakumar, "Pellagra in the United States"; Bollet, "Politics and Pellagra"; Goldberger et al., *The Experimental Production of Pellagra in Human Subjects by Means of Diet*; Mariani-Costantini and Mariani-Costantini, "An Outline of the History of Pellagra in Italy"; Elmore and Feinstein, "Joseph Goldberger."

5. Rajakumar, "Pellagra in the United States"; Bollet, "Politics and Pellagra"; Goldberger et al., *The Experimental Production of Pellagra in Human Subjects by Means of Diet*; Elmore and Feinstein, "Joseph Goldberger."

6. Whitaker, "Bread and Work"; Livi-Bacci, "Fertility, Nutrition, and Pellagra."

7. Katz, Hediger, and Valleroy, "Traditional Maize Processing Techniques in the New World"; Wall and Carpenter, "Variation in Availability of Niacin in Grain Products"; Rajakumar, "Pellagra in the United States"; Bollet, "Politics and Pellagra"; Goldberger et al., *The Experimental Production of Pellagra in Human Subjects by Means of Diet.*

8. Barnes, *Diseases and Human Evolution.*

9. Drummond and Wilbraham, *The Englishman's Food.*

10. Weick, "A History of Rickets in the United States."

11. Guallar et al., "Enough Is Enough."

12. Among 400,000 young Singaporean men called up for compulsory preenlistment medical screening, the prevalence of myopia (defined as unaided visual acuity worse than 6/18) increased from 26.3 percent during 1974–84 to 43.3% during 1987–91. Angle and Wissmann, "The Epidemiology of Myopia"; Brown, "Use-Abuse Theory of Changes in Refraction Versus Biologic Theory"; Rose et al., "Outdoor Activity Reduces the Prevalence of Myopia in Children"; Saw, "A Synopsis of the Prevalence Rates and Environmental Risk Factors for Myopia"; Au Eong, Tay, and Lim, "Education and Myopia in 110,236 Young Singapore Males"; Tay et al., "Myopia and Educational Attainment in 421,116 Young Singaporean Males."

13. Jones et al., "Parental History of Myopia, Sports and Outdoor Activities, and Future Myopia"; Rose et al., "Outdoor Activity Reduces the Prevalence of Myopia in Children"; Dirani et al., "Outdoor Activity and Myopia in Singapore Teenage Children."

14. Dirani et al., "Outdoor Activity and Myopia in Singapore Teenage Children"; Jones et al., "Parental History of Myopia, Sports and Outdoor Activities, and Future Myopia"; Smith, Hung, and Huang, "Protective Effects of High Ambient Lighting on the Development of Form-Deprivation Myopia in Rhesus Monkeys"; Ashby, Ohlendorf, and Schaeffel, "The Effect of Ambient Illuminance on the Development of Deprivation Myopia in Chicks"; Fujiwara et al., "Seasonal Variation in Myopia Progression and Axial Elongation"; Meng et al., "Myopia and Iris Colour"; Sherwin et al., "The Association Between Time Spent Outdoors and Myopia Using a Novel Biomarker of Outdoor Light Exposure."

15. Beauchemin and Hays, "Sunny Hospital Rooms Expedite Recovery from Severe and Refractory Depressions"; Beauchemin and Hays, "Dying in the Dark."

16. Kinney et al., "Relation of Schizophrenia Prevalence to Latitude, Climate, Fish Consumption, Infant Mortality, and Skin Color"; Saha et al., "The Incidence and Prevalence of Schizophrenia Varies with Latitude"; Grant and Soles, "Epidemiologic Evidence for Supporting the Role of Maternal Vitamin D Deficiency as a Risk Factor for the Development of Infantile Autism."

17. Parra, "Human Pigmentation Variation"; Norton et al., "Genetic Evidence for the Convergent Evolution of Light Skin in Europeans and East Asians."

18. Gandini et al., "Meta-Analysis of Risk Factors for Cutaneous Melanoma"; Elwood and Jopson, "Melanoma and Sun Exposure"; Westerdahl et al., "Sunscreen Use and Malignant Melanoma"; Green et al., "Reduced Melanoma After Regular Sunscreen Use"; Bastuji-Garin and Diepgen, "Cutaneous Malignant Melanoma, Sun Exposure, and Sunscreen Use"; de Gruijl, "Skin Cancer

and Solar UV Radiation"; Holick, "Environmental Factors That Influence the Cutaneous Production of Vitamin D."

19. In the study, 2,848 infants were submitted to skin-prick and food-ingestion tests. The one-year-old infants were not a completely random sample from the general Melbourne population of one-year-olds, because parents who agreed to participate tended to come from higher-income families, and their children tended to have prior eczema conditions. Prescott and Allen, "Food Allergy"; Osborne et al., "Prevalence of Challenge-Proven IgE-Mediated Food Allergy Using Population-Based Sampling and Predetermined Challenge Criteria in Infants."

20. Shek and Lee, "Food Allergy in Asia."

21. Ninety-eight pregnant women who were booked for delivery in Western Australia and had a history of allergies were allotted four 1-gram fish oil pills a day.

22. The studies were conducted in Sweden and Norway. In Sweden, children who ate more fish in early life had lower risks of asthma, eczema, and allergic rhinitis and produced fewer antibodies in an allergen blood test at four years of age. In Norway, children who ate fish during their first year of life had lower rates of hay fever, again at four years of age. The Swedish study involved 4,089 children, the Norwegian study 2,531 children.

23. The survey involved 691 Southern California public school children from fourth to tenth grade. Oily fish were defined as blue mackerel, Atlantic salmon, southern bluefin tuna, blue-eye trevalla, rainbow trout, mullet, blue grenadier, tailor, silver bream, gemfish, blackfish, orange roughy, pilchards, redfish, yellowtail, and tarwhine.

24. Anandan, Nurmatov, and Sheikh, "Omega 3 and 6 Oils for Primary Prevention of Allergic Disease"; Dunstan et al., "Fish Oil Supplementation in Pregnancy Modifies Neonatal Allergen-Specific Immune Responses and Clinical Outcomes in Infants at High Risk of Atopy"; Kull et al., "Fish Consumption During the First Year of Life and Development of Allergic Diseases During Childhood"; Thien, Mencia-Huerta, and Lee, "Dietary Fish Oil Effects on Seasonal Hay Fever and Asthma in Pollen-Sensitive Subjects"; Nafstad et al., "Asthma and Allergic Rhinitis at 4 Years of Age in Relation to Fish Consumption in Infancy"; Salam, Li, Langholz, and Gilliland. "Maternal Fish Consumption During Pregnancy and Risk of Early Childhood Asthma"; Sausenthaler et al., "Maternal Diet During Pregnancy in Relation to Eczema and Allergic Sensitization in the Offspring at 2 Y of Age."

25. Anandan, Nurmatov, and Sheikh, "Omega 3 and 6 Oils for Primary Prevention of Allergic Disease"; Moher et al., *Health Effects of Omega-3 Fatty Acids on Asthma.*

26. Holick, "Vitamin D Deficiency"; Van Belle, Gysemans, and Mathieu, "Vitamin D in Autoimmune, Infectious and Allergic Diseases: A Vital Player?"; Vassallo and Camargo Jr., "Potential Mechanisms for the Hypothesized Link Between Sunshine, Vitamin D, and Food Allergy in Children."

27. Devereux et al., "Maternal Vitamin D Intake During Pregnancy and Early Childhood Wheezing"; Gupta et al., "Vitamin D and Asthma in Children."

28. Researchers surveyed 1,669 mothers. Allergic rhinitis and asthma were assessed at five years of age. Vitamin D in food came principally from fish and margarine. This study was conducted among children at risk of developing type 1 diabetes, which may offer protection against allergic diseases. However, the rates of asthma and allergic rhinitis among infants in this study were similar to rates in the general Finnish population. Also, the effect of vitamin D was consistent with two previously mentioned studies, conducted in North America and Scotland.

29. Erkkola et al., "Maternal Vitamin D Intake During Pregnancy Is Inversely Associated with Asthma and Allergic Rhinitis in 5-Year-Old Children"; Wjst and Hyppönen, "Vitamin D Serum Levels and Allergic Rhinitis."

30. Doctors checked American data on 1,511,534 EpiPen prescriptions filled in 2004. Massachusetts had 11.8 EpiPen prescriptions per 1,000 people. Hawaii had 2.7 EpiPen prescriptions per 1,000.

31. Simons, Peterson, and Black, "Epinephrine Dispensing Patterns for an Out-of-Hospital Population"; Camargo et al., "Regional Differences in EpiPen Prescriptions in the United States"; Mullins, Clark, and Camargo, "Regional Variation in Epinephrine Autoinjector Prescriptions in Australia."

32. The hypoallergenic baby formula study was conducted in 2010. The peanut and egg allergy study was conducted in 2012. Eight-year-old and nine-year-old kids living in northern (i.e., colder) states were more likely to develop peanut allergies. Kids between four and five years of age followed the same pattern, with the addition of egg allergies.

33. Mullins, Clark, and Camargo, "Regional Variation in Infant Hypoallergenic Formula Prescriptions in Australia"; Rudders, Espinola, and Camargo, "North-South Differences in US Emergency Department Visits for Acute Allergic Reactions"; Vassallo et al., "Season of Birth and Food Allergy in Children"; Osborne et al., "Prevalence of Eczema and Food Allergy Is Associated with Latitude in Australia."

34. In the Boston study, a group of eleven children suffering from wintertime eczema took either vitamin D or an identical-looking placebo. Four out of five kids who took 1000 IU of vitamin D daily showed improvement in their eczema symptoms; only one among the six kids who got the placebo showed

improvement. The study in Iran was a randomized control study of fifty-two teens and adults. The Italian study involved thirty-seven children. Vocks, "Climatotherapy in Atopic Eczema"; Byremo, Rød, and Carlsen, "Effect of Climatic Change in Children with Atopic Eczema"; Harari et al., "Climatotherapy of Atopic Dermatitis at the Dead Sea"; Sidbury et al., "Randomized Controlled Trial of Vitamin D Supplementation for Winter-Related Atopic Dermatitis in Boston"; Javanbakht et al., "Randomized Controlled Trial Using Vitamins E and D Supplementation in Atopic Dermatitis"; Peroni et al., "Correlation Between Serum 25-Hydroxyvitamin D Levels and Severity of Atopic Dermatitis in Children."

35. Hata et al., "Administration of Oral Vitamin D Induces Cathelicidin Production in Atopic Individuals"; Meyer and Thyssen, "Filaggrin Gene Defects and Dry Skin Barrier Function"; Osawa et al., "Japanese-Specific Filaggrin Gene Mutations in Japanese Patients Suffering from Atopic Eczema and Asthma"; Chen et al., "Wide Spectrum of Filaggrin-Null Mutations in Atopic Dermatitis Highlights Differences Between Singaporean Chinese and European Populations."

36. Wjst and Hyppönen, "Vitamin D Serum Levels and Allergic Rhinitis"; Hypponen, "Infant Vitamin D Supplementation and Allergic Conditions in Adulthood"; Bäck et al., "Does Vitamin D Intake During Infancy Promote the Development of Atopic Allergy?"; Gale et al., "Maternal Vitamin D Status During Pregnancy and Child Outcomes"; Ahn et al., "Serum Vitamin D Concentration and Prostate Cancer Risk"; Chen et al., "Prospective Study of Serum 25(OH)-Vitamin D Concentration and Risk of Oesophageal and Gastric Cancers"; Abnet et al., "Serum 25(OH)-Vitamin D Concentration and Risk of Esophageal Squamous Dysplasia"; Fox, "Frank C. Garland, 60, Who Connected Vitamin D Deficiency and Cancer, Dies."

37. Rowland et al., "Calcium Intake and Prostate Cancer Among African Americans"; Gupta et al., "Vitamin D and Asthma in Children"; Vassallo and Camargo , "Potential Mechanisms for the Hypothesized Link Between Sunshine, Vitamin D, and Food Allergy in Children"; Grady et al., "Hormone Therapy to Prevent Disease and Prolong Life in Postmenopausal Women"; Grady et al., "Cardiovascular Disease Outcomes During 6.8 Years of Hormone Therapy"; Guallar et al., "Postmenopausal Hormone Therapy"; Ravdin et al., "The Decrease in Breast-Cancer Incidence in 2003 in the United States"; Hawkes et al., "Grandmothering, Menopause, and the Evolution of Human Life Histories"; Liu et al., "Systematic Review"; Bhasin et al., "The Effects of Supraphysiologic Doses of Testosterone on Muscle Size and Strength in Normal Men"; Nieminen

et al., "Serious Cardiovascular Side Effects of Large Doses of Anabolic Steroids in Weight Lifters."

38. Waite, "Blackley and the Development of Hay Fever as a Disease of Civilization in the Nineteenth Century."

39. Strachan, "Hay Fever, Hygiene, and Household Size."

40. Anyo et al., "Early, Current and Past Pet Ownership"; von Mutius, "99th Dahlem Conference on Infection, Inflammation and Chronic Inflammatory Disorders"; Schaub, Lauener, and von Mutius, "The Many Faces of the Hygiene Hypothesis"; Cooper, "Intestinal Worms and Human Allergy"; Cooper, "Interactions Between Helminth Parasites and Allergy"; Figueiredo et al., "Chronic Intestinal Helminth Infections Are Associated with Immune Hyporesponsiveness and Induction of a Regulatory Network"; Bloomfield et al., "Too Clean, or Not Too Clean"; Sherriff and Golding, "Hygiene Levels in a Contemporary Population Cohort Are Associated with Wheezing and Atopic Eczema in Preschool Infants"; Berdoy, Webster, and Macdonald, "Fatal Attraction in Rats Infected with Toxoplasma Gondii"; Zhang et al., "*Toxoplasma gondii* Immunoglobulin G Antibodies and Nonfatal Suicidal Self-Directed Violence"; Blaser, *Missing Microbes.*

41. Barnes, *Diseases and Human Evolution.*

42. Barnes, *Diseases and Human Evolution.*

43. Including sickle cell anemia, hemoglobin variants C and E, the Duffy Negative Blood Group, thalassemia syndromes, and the G6PD enzyme defect.

44. Barnes, *Diseases and Human Evolution.*

45. Barnes, *Diseases and Human Evolution.*

46. Gire et al., "Genomic Surveillance Elucidates Ebola Virus Origin and Transmission During the 2014 Outbreak"; Vogel, "Genomes Reveal Start of Ebola Outbreak"; Li and Chen, "Evolutionary History of Ebola Virus"; Barnes, *Diseases and Human Evolution.*

47. Wong, Bundy, and Golden, "The Rate of Ingestion of *Ascaris lumbricoides* and *Trichuris trichiura* Eggs in Soil and Its Relationship to Infection in Two Children's Homes in Jamaica"; Gutiérrez, *Diagnostic Pathology of Parasitic Infections*; ICDDR, *Diarrhoeal Diseases Research*; Qian, *Nematode Nicotinic Acetylcholine Receptors*; Read and Skorping, "The Evolution of Tissue Migration by Parasitic Nematode Larvae"; Mulcahy et al., "Tissue Migration by Parasitic Helminths—An Immuno-evasive Strategy?"

48. Wong, Bundy, and Golden, "The Rate of Ingestion of *Ascaris lumbricoides* and *Trichuris trichiura* Eggs in Soil and Its Relationship to Infection in Two Children's Homes in Jamaica"; Gutiérrez, *Diagnostic Pathology of Parasitic Infections*; Qian,

*Nematode Nicotinic Acetylcholine Receptors*; Read and Skorping, "The Evolution of Tissue Migration by Parasitic Nematode Larvae"; Mulcahy et al., "Tissue Migration by Parasitic Helminths—An Immunoevasive Strategy?"

49. Gutiérrez, *Diagnostic Pathology of Parasitic Infections*; ICDDR, *Diarrhoeal Diseases Research*; Fernando, Fernando, and Leong, *Tropical Infectious Diseases*.

50. Figueiredo et al., "Chronic Intestinal Helminth Infections Are Associated with Immune Hyporesponsiveness and Induction of a Regulatory Network."

51. Pearce et al., "Worldwide Trends in the Prevalence of Asthma Symptoms."

52. Cooper, "Interactions Between Helminth Parasites and Allergy"; Bloomfield et al., "Too Clean, or Not Too Clean."

53. Summers et al., "*Trichuris suis* Therapy in Crohn's Disease"; Summers et al., "*Trichuris suis* Therapy for Active Ulcerative Colitis"; Laskaris, *Color Atlas of Oral Diseases*; DiMarino and Benjamin, *Gastrointestinal Disease*; Bloch, "Could Kashrut Be Partly to Blame for Crohn's Disease?"; Weinstock and Elliott, "Translatability of Helminth Therapy in Inflammatory Bowel Diseases."

54. Klugman et al., "A Trial of a 9-Valent Pneumococcal Conjugate Vaccine in Children with and Those Without HIV Infection"; Silverberg et al., "Chickenpox in Childhood Is Associated with Decreased Atopic Disorders, IgE, Allergic Sensitization, and Leukocyte Subsets."

55. Summers et al., "*Trichuris suis* Therapy for Active Ulcerative Colitis"; Correale and Farez, "Association Between Parasite Infection and Immune Responses in Multiple Sclerosis"; Saunders et al., "Inhibition of Autoimmune Type 1 Diabetes by Gastrointestinal Helminth Infection"; Adams, "Gut Instinct"; DeLong, "Conflicts of Interest in Vaccine Safety Research."

## THE CALORIE CONUNDRUM

1. USDA Economic Research Service, "Food Expenditures."

2. Zhou et al., "Nutrient Intakes of Middle-Aged Men and Women in China, Japan, United Kingdom, and United States in the Late 1990s."

3. Turner, "The Calorie Restriction Dieters."

4. Nakagawa et al., "Comparative and Meta-Analytic Insights into Life Extension via Dietary Restriction."

5. Renehan, "Insulin-like Growth Factor (IGF)-I, IGF Binding Protein-3, and Cancer Risk"; Juul, "Serum Levels of Insulin-like Growth Factor I and Its Binding Proteins in Health and Disease."

6. Nakagawa et al., "Comparative and Meta-Analytic Insights into Life Extension via Dietary Restriction."

7. Shanley and Kirkwood, "Calorie Restriction and Aging."

8. "Glossary of Sexual and Scatological Euphemisms"; Vitousek, "Caloric Restriction for Longevity."

9. Mattison et al., "Impact of Caloric Restriction on Health and Survival in Rhesus Monkeys from the NIA Study."

10. Lawler et al., "Diet Restriction and Ageing in the Dog."

11. Willcox, Willcox, and Suzuki, *The Okinawa Diet Plan*; Zhou et al., "Nutrient Intakes of Middle-Aged Men and Women in China, Japan, United Kingdom, and United States in the Late 1990s"; Fontana, "Long-Term Effects of Calorie or Protein Restriction on Serum IGF-1 and IGFBP-3 Concentration in Humans."

12. Leigh, *The World's Greatest Fix*.

13. Preceding millet with, for example, green gram beans, adzuki beans, cucurbits, cannabis or hemp, sesame, rape, and soybean.

14. Leigh, *The World's Greatest Fix*, gives the population "in England within the Norman domains in 1086" as 283,242 males (women and children were not counted).

15. The Chinese were aware of some life-supporting agent in the air, noted as the yin of the air in the eighth century AD, but beyond this there was no deep understanding of why their assorted agricultural techniques worked.

16. Dense agricultural civilizations also existed in South Asia and the Middle East, but the climates in these regions promoted the spread of destructive infectious diseases such as malaria.

17. The results of the experiment were published on May 4, 1692.

18. May, *World Population Policies*; Grinin, De Munck, and Korotaev, *History and Mathematics*.

19. Leigh, *The World's Greatest Fix*.

20. Newcomb and Spurr, *A Technical History of the Motor Car*.

21. Zhou et al., "Nutrient Intakes of Middle-Aged Men and Women in China, Japan, United Kingdom, and United States in the Late 1990s."

22. Jenike, "Nutritional Ecology."

23. Dugas et al., "Energy Expenditure in Adults Living in Developing Compared with Industrialized Countries."

24. The physical activity level of people in countries ranked low and middle on the Human Development Index or who work at farming and in factories is higher.

25. BMI is calculated as weight divided by height, specifically: kg/m2, or lb/in2×703.

26. As of 2010. "FastStats: Body Measurements."

27. de Garine and Koppert, "*Guru*-Fattening Sessions Among the Massa."

28. Brink, "The Fattening Room Among the Annang of Nigeria."

29. Mattson and Wan, "Beneficial Effects of Intermittent Fasting and Caloric Restriction on the Cardiovascular and Cerebrovascular Systems"; Chausse et al., "Intermittent Fasting Induces Hypothalamic Modifications Resulting in Low Feeding Efficiency, Low Body Mass and Overeating"; Barnosky et al., "Intermittent Fasting vs Daily Calorie Restriction for Type 2 Diabetes Prevention"; Cerqueira and Kowaltowski, "Mitochondrial Metabolism in Aging."

30. Cerqueira et al., "Long-Term Intermittent Feeding, but Not Caloric Restriction, Leads to Redox Imbalance, Insulin Receptor Nitration, and Glucose Intolerance."

31. Trepanowski and Bloomer, "The Impact of Religious Fasting on Human Health."

32. Sadeghirad et al., "Islamic Fasting and Weight Loss."

33. Westerterp and Speakman, "Physical Activity Energy Expenditure Has Not Declined since the 1980s and Matches Energy Expenditures of Wild Mammals."

34. Taubes, *Good Calories, Bad Calories*.

35. Marlowe, "Hunter-Gatherers and Human Evolution."

36. Hu et al., "Television Watching and Other Sedentary Behaviors in Relation to Risk of Obesity and Type 2 Diabetes Mellitus in Women"; Grøntved and Hu, "Television Viewing and Risk of Type 2 Diabetes, Cardiovascular Disease, and All-Cause Mortality"; Nielsen, *State of the Media TV Usage Trends: Q2 2010*.

37. Hu et al., "Television Watching and Other Sedentary Behaviors in Relation to Risk of Obesity and Type 2 Diabetes Mellitus in Women"; Grøntved and Hu, "Television Viewing and Risk of Type 2 Diabetes, Cardiovascular Disease, and All-Cause Mortality"; Nielsen, *State of the Media TV Usage Trends: Q2 2010*..

38. Sugiyama, Ding, and Owen, "Commuting by Car."

39. Sieber et al., "Obesity and Other Risk Factors."

40. Scholey, Harper, and Kennedy, "Cognitive Demand and Blood Glucose"; Fairclough and Houston, "A Metabolic Measure of Mental Effort."

41. Miller and Bender, "The Breakfast Effect."

42. Lund et al., "Prevalence and Risk Factors for Obesity in Adult Cats from Private US Veterinary Practices"; McGreevy et al., "Prevalence of Obesity in Dogs Examined by Australian Veterinary Practices and the Risk Factors Involved."

43. Trasande et al., "Infant Antibiotic Exposures and Early-Life Body Mass."

44. Flegal et al., "Association of All-Cause Mortality with Overweight and Obesity Using Standard Body Mass Index Categories."

45. Dixon et al., "'Obesity Paradox' Misunderstands the Biology of Optimal Weight Throughout the Life Cycle."

46. Jacobi and Cash, "In Pursuit of the Perfect Appearance"; Frederick, Fessler, and Haselton, "Do Representations of Male Muscularity Differ in Men's and Women's Magazines?"

47. Frederick, Fessler, and Haselton, "Do Representations of Male Muscularity Differ in Men's and Women's Magazines?"

48. Allbaugh, *Crete*.

49. The survey by Hatzis et al. in 2010 considered men between the ages of 53 and 73, while the original survey by Keys et al. considered younger men, between the ages of 40 and 59; thus the figures of nutrient intake can only be compared for approximate differences in magnitudes.

50. Vardavas, *Public Health Implications of the Mediterranean Diet*.

51. Lionis et al., "A High Prevalence of Diabetes Mellitus in a Municipality of Rural Crete, Greece."

52. Hatzis et al., "A 50-Year Follow-up of the Seven Countries Study."

53. Vardavas, *Public Health Implications of the Mediterranean Diet*; Hatzis et al., "A 50-Year Follow-up of the Seven Countries Study."

54. Tambalis et al., "Higher Prevalence of Obesity in Greek Children Living in Rural Areas Despite Increased Levels of Physical Activity."

55. Willcox et al., "Caloric Restriction, the Traditional Okinawan Diet, and Healthy Aging"; Le Bourg, "About the Article 'Exploring the Impact of Climate on Human Longevity' (Exp. Geront. 47, 660-671, 2012)."

56. Suzuki, "The Okinawa Shock."

57. Inoue, *Okinawa and the U.S. Military*; Molasky, *The American Occupation of Japan and Okinawa*; Murray, *Atlas of American Military History*.

58. Takasu et al., "Influence of Motorization and Supermarket-Proliferation on the Prevalence of Type 2 Diabetes in the Inhabitants of a Small Town on Okinawa, Japan"; Joyce, "Japanese Get a Taste for Western Food and Fall Victim to Obesity and Early Death"; Suzuki, "The Okinawa Shock"; Todoriki, Willcox, and Willcox, "The Effects of Post-War Dietary Change on Longevity and Health in Okinawa."

59. Higgins, "Epidemiology of Constipation in North America."

60. Sikirov, "Comparison of Straining During Defecation in Three Positions"; Sakakibara et al., "Influence of Body Position on Defecation in Humans."

61. Chakrabarti et al., "Is Squatting a Triggering Factor for Stroke in Indians?"

## THE FUTURE OF FOOD

1. Serra-Majem et al., "How Could Changes in Diet Explain Changes in Coronary Heart Disease Mortality in Spain?"; Fried and Rao, "Sugars, Hypertriglyceridemia, and Cardiovascular Disease."

2. Rogers et al., "Diet Throughout Childhood and Age at Menarche in a Contemporary Cohort of British Girls"; Tehrani et al., "Intake of Dairy Products, Calcium, Magnesium, and Phosphorus in Childhood and Age at Menarche in the Tehran Lipid and Glucose Study"; Tamakoshi, Yatsuya, and Tamakoshi, "Early Age at Menarche Associated with Increased All-Cause Mortality"; Frisch, *Female Fertility and the Body Fat Connection.*

3. "Leader in Healthcare and Preventive Medicine: Dean Ornish, MD."

4. Fallon and Enig, *Nourishing Traditions.*

5. "Turning the Food Pyramid on Its Head with Sally Fallon Morrell."

6. "State-by-State Review of Raw Milk Laws."

7. Weston A. Price Foundation, "Journal, Summer 2013, Our Broken Food System."

8. Price-Pottenger Nutrition Foundation, "Traditional Diets."

9. Willcox, Willcox, and Suzuki, *The Okinawa Diet Plan*; Zabilka, *Customs and Cultures of Okinawa*; Kerr, *Okinawa: The History of an Island People.*

10. Buettner, *The Blue Zones.*

11. Buettner, *The Blue Zones*; Poulain et al., "Identification of a Geographic Area Characterized by Extreme Longevity in the Sardinia Island."

12. Fernando and Hill, *Lentil as Anything.*

13. Busfield et al., "A Genomewide Search for Type 2 Diabetes–Susceptibility Genes in Indigenous Australians."

14. Adams, "Sportsman's Shot, Poacher's Pot"; Mahoney, "Recreational Hunting and Sustainable Wildlife Use in North America."

15. Mahoney, "Recreational Hunting and Sustainable Wildlife Use in North America."

16. Howard, "Salmon Farming Gets Leaner and Greener"; Knapp, Roheim, and Anderson, *The Great Salmon Run.*

17. "About Cooke Aquaculture."

18. Harvey and Milewski, *Salmon Aquaculture in the Bay of Fundy.*

19. Worm and Branch, "The Future of Fish."

20. Domingo and Giné Bordonaba, "A Literature Review on the Safety Assessment of Genetically Modified Plants."

21. Lesser et al., "Relationship Between Funding Source and Conclusion Among Nutrition-Related Scientific Articles"; Diels et al., "Association of Financial or Professional Conflict of Interest to Research Outcomes on Health Risks or Nutritional Assessment Studies of Genetically Modified Products."

22. De Vendômois et al., "A Comparison of the Effects of Three GM Corn Varieties on Mammalian Health"; Cisterna et al., "Can a Genetically Modified Organism–Containing Diet Influence Embryo Development?"

23. "USA: Cultivation of GM Plants, 2013."

24. Kopicki, "Strong Support for Labeling Modified Foods."

25. Wilson, "Maine Becomes Second State to Require GMO Labels"; Reilly, "Malloy Signs State GMO Labeling Law in Fairfield."

26. Hallenbeck, "Vermont Defends GMO Labeling Law."

27. "Ethanol/Corn Balance Sheets—Agricultural Marketing Resource Center."

28. Miller et al., "Health Status, Health Conditions, and Health Behaviors Among Amish Women"; Bassett, Schneider, and Huntington, "Physical Activity in an Old Order Amish Community"; Stevick, *Growing Up Amish.*

## AFTERWORD: RULES TO EAT AND LIVE BY

1. Hehemann et al., "Transfer of Carbohydrate-Active Enzymes from Marine Bacteria to Japanese Gut Microbiota"; Perry et al., "Diet and the Evolution of Human Amylase Gene Copy Number Variation"; Luca, Perry, and Di Rienzo, "Evolutionary Adaptations to Dietary Changes"; Falchi et al., "Low Copy Number of the Salivary Amylase Gene Predisposes to Obesity."

2. Ou et al., "Excess Winter Mortality and Cold Temperatures in a Subtropical City, Guangzhou, China."

3. Blaser, *Missing Microbes.*

4. Uribarri et al., "Advanced Glycation End Products in Foods and a Practical Guide to Their Reduction in the Diet"; Semba et al., "Advanced Glycation End Products and Their Circulating Receptors Predict Cardiovascular Disease Mortality in Older Community-Dwelling Women"; Semba, Nicklett, and Ferrucci, "Does Accumulation of Advanced Glycation End Products Contribute to the Aging Phenotype?"; Vlassara and Striker, "The Role of Advanced Glycation End-Products in the Etiology of Insulin Resistance and Diabetes."

5. Uribarri et al., "Advanced Glycation End Products in Foods and a Practical Guide to Their Reduction in the Diet"; Semba, Nicklett, and Ferrucci, "Does Accumulation of Advanced Glycation End Products Contribute to the Aging Phenotype?"

# BIBLIOGRAPHY

Abbasi, A. A., A. S. Prasad, P. Rabbani, and E. DuMouchelle. "Experimental Zinc Deficiency in Man: Effect on Testicular Function." *Journal of Laboratory and Clinical Medicine* 96, no. 3 (1980): 544–50.

Abdelgadir, Salaheldin E., A. G. A. Wahbi, and O. F. Idris. "Some Blood and Plasma Constituents of the Camel." In *The Camelid: An All-Purpose Animal,* edited by Ross Cockrill, 438–43. Scandinavian Institute of African Studies, 1979.

Abnet, Christian C., Wen Chen, Sanford M. Dawsey, Wen-Qiang Wei, Mark J. Roth, Bing Liu, Ning Lu, Philip R. Taylor, and You-Lin Qiao. "Serum 25(OH)-Vitamin D Concentration and Risk of Esophageal Squamous Dysplasia." *Cancer Epidemiology Biomarkers and Prevention* 16, no. 9 (September 1, 2007): 1889–93. doi:10.1158/1055-9965.EPI-07-0461.

"About Cooke Aquaculture." Accessed August 28, 2014. www.cookeaqua.com /index.php/about-cooke-aquaculture.

Adams, Tim. "Gut Instinct: The Miracle of the Parasitic Hookworm." *Guardian,* May 23, 2010. www.guardian.co.uk/lifeandstyle/2010/may/23/parasitic-hook worm-jasper-lawrence-tim-adams.

Adams, William M. "Sportsman's Shot, Poacher's Pot: Hunting, Local People and the History of Conservation." In *Recreational Hunting, Conservation and Rural Livelihoods,* edited by Barney Dickson, Jon Hutton, and William M. Adams, 125–40. Wiley-Blackwell, 2009. http://onlinelibrary.wiley.com/doi/10.1002/9781 444303179.ch8/summary.

Ahn, Jiyoung, Ulrike Peters, Demetrius Albanes, Mark P. Purdue, Christian C. Abnet, Nilanjan Chatterjee, Ronald L. Horst, Bruce W. Hollis, Wen-Yi Huang, James M. Shikany, and Richard B. Hayes. "Serum Vitamin D Concentration and Prostate Cancer Risk: A Nested Case-Control Study." *Journal of the National Cancer Institute* 100, no. 11 (June 4, 2008): 796–804. doi:10.1093/jnci/djn152.

Akazawa, Yoshihiro, Shoji Watanabe, Shigenori Nobukiyo, Hiroya Iwatake, Yoshitake Seki, Tsuyoshi Umehara, Kouichiro Tsutsumi, and Izumi Koizuka. "The Management of Possible Fishbone Ingestion." *Auris Nasus Larynx* 31, no. 4 (December 2004): 413–16. doi:10.1016/j.anl.2004.09.007.

Alcock, Joan Pilsbury. *Food in the Ancient World.* Greenwood Publishing Group, 2006.

Aldemir, M., E. Okulu, S. Neşelioğlu, O. Erel, and Ö Kayıgil. "Pistachio Diet Improves Erectile Function Parameters and Serum Lipid Profiles in Patients with Erectile Dysfunction." *International Journal of Impotence Research* 23, no. 1 (2011): 32–38.

Alinia, Sevil, O. Hels, and I. Tetens. "The Potential Association Between Fruit Intake and Body Weight—A Review." *Obesity Reviews* 10, no. 6 (2009): 639–47.

Allbaugh, Leland G. *Crete: A Case Study of an Underdeveloped Area.* Princeton University Press, 1953.

Allport, Susan. *The Queen of Fats: Why Omega-3s Were Removed from the Western Diet and What We Can Do to Replace Them.* University of California Press, 2008.

Altuğ, Tomris. *Introduction to Toxicology and Food: Toxin Science, Food Toxicants, Chemoprevention.* CRC Press, 2003.

Anandan, C., U. Nurmatov, and A. Sheikh. "Omega 3 and 6 Oils for Primary Prevention of Allergic Disease: Systematic Review and Meta-Analysis." *Allergy* 64, no. 6 (2009): 840–48. doi:10.1111/j.1398-9995.2009.02042.x.

Anderson, A., C. Sand, F. Petchey, and T. H. Worthy. "Faunal Extinction and Human Habitation in New Caledonia: Initial Results and Implications of New Research at the Pindai Caves." *Journal of Pacific Archaeology* 1, no. 1 (2010): 89–109.

Angle, John, and David Wissmann. "The Epidemiology of Myopia." *American Journal of Epidemiology* 111, no. 2 (February 1, 1980): 220–28.

Antón, S. "Natural History of *Homo erectus.*" *Yearbook of Physical Anthropology* 46 (2003): 126–70.

Anyo, G., B. Brunekreef, G. De Meer, F. Aarts, N. A. H. Janssen, and P. Van Vliet. "Early, Current and Past Pet Ownership: Associations with Sensitization, Bronchial Responsiveness and Allergic Symptoms in School Children." *Clinical and Experimental Allergy* 32, no. 3 (2002): 361–66. doi:10.1046/j.1365-2222.2002 .01254.x.

Arditti, Joseph, and Eloy Rodriguez. "Dieffenbachia: Uses, Abuses and Toxic Constituents: A Review." *Journal of Ethnopharmacology* 5, no. 3 (May 1982): 293–302. doi:10.1016/0378-8741(82)90015-0.

Arnoldi, Anna. *Functional Foods, Cardiovascular Disease, and Diabetes.* Woodhead Publishing, 2004.

Ashby, Regan, Arne Ohlendorf, and Frank Schaeffel. "The Effect of Ambient Illuminance on the Development of Deprivation Myopia in Chicks." *Investigative Ophthalmology and Visual Science* 50, no. 11 (November 1, 2009): 5348–54. doi:10.1167/iovs.09-3419.

Au Eong, K. G., T. H. Tay, and M. K. Lim. "Education and Myopia in 110,236 Young Singapore Males." *Singapore Medical Journal* 34, no. 6 (1993): 489–92.

Ayyagari, Radha, B. S. Narasinga Rao, and D. N. Roy. "Lectins, Trypsin Inhibitors, BOAA and Tannins in Legumes and Cereals and the Effects of Processing." *Food Chemistry* 34, no. 3 (1989): 229–38. doi:10.1016/0308-8146(89)90143-X.

Bäck, Ove, Hans Blomquist, Olle Hernell, and Berndt Stenberg. "Does Vitamin D Intake During Infancy Promote the Development of Atopic Allergy?" *Acta Dermato-Venereologica* 89, no. 1 (2009): 28–32. doi:10.2340/00015555-0541.

Bainbridge, D. A. "The Rise of Agriculture: A New Perspective." *Ambio* 14, no. 3 (1985): 148–51.

———. "The Use of Acorns for Food in California: Past, Present, Future." In *Symposium on Multiple-Use Management of California's Hardwoods*, 453–58. USDA Pacific Southwest Forest and Range Experiment Station, 1987. www.fs.fed.us /psw/publications/documents/psw_gtr100/psw_gtr100a.pdf.

Balny, Claude, Patrick Masson, and K. Heremans. *Frontiers in High Pressure Biochemistry and Biophysics.* Elsevier, 2002.

Barceloux, Donald G. *Medical Toxicology of Natural Substances: Foods, Fungi, Medicinal Herbs, Plants, and Venomous Animals.* John Wiley & Sons, 2012.

Bar-Yosef, Ofer. "Climatic Fluctuations and Early Farming in West and East Asia." *Current Anthropology* 52, no. S4 (2011): S175–93.

Barnes, Ethne. *Diseases and Human Evolution.* University of New Mexico Press, 2005.

Barnosky, Adrienne R., Kristin K. Hoddy, Terry G. Unterman, and Krista A. Varady. "Intermittent Fasting vs Daily Calorie Restriction for Type 2 Diabetes Prevention: A Review of Human Findings." *Translational Research* 164, no. 4 (October 2014): 302–11. www.sciencedirect.com/science/article/pii /S193152441400200X.

Barona, Jacqueline, and Maria Luz Fernandez. "Dietary Cholesterol Affects Plasma Lipid Levels, the Intravascular Processing of Lipoproteins and Reverse Cholesterol Transport Without Increasing the Risk for Heart Disease." *Nutrients* 4, no. 12 (August 17, 2012): 1015–25. doi:10.3390/nu4081015.

Bassett, David R., Jr., Patrick L. Schneider, and Gertrude E. Huntington. "Physical Activity in an Old Order Amish Community." *Medicine and Science in Sports and Exercise* 36, no. 1 (2004): 79–85.

Bastuji-Garin, S., and T. L. Diepgen. "Cutaneous Malignant Melanoma, Sun Exposure, and Sunscreen Use: Epidemiological Evidence." *British Journal of Dermatology* 146, no. S61 (2002): 24–30. doi:10.1046/j.1365-2133.146.s61.9.x.

Beauchemin, K. M., and P. Hays. "Dying in the Dark: Sunshine, Gender and Outcomes in Myocardial Infarction." *Journal of the Royal Society of Medicine* 91, no. 7 (July 1998): 352–54.

———. "Sunny Hospital Rooms Expedite Recovery from Severe and Refractory Depressions." *Journal of Affective Disorders* 40, nos. 1–2 (September 9, 1996): 49–51. doi:10.1016/0165-0327(96)00040-7.

Becker, Michael A., H. Ralph Schumacher Jr, Robert L. Wortmann, Patricia A. MacDonald, Denise Eustace, William A. Palo, Janet Streit, and Nancy Joseph-Ridge. "Febuxostat Compared with Allopurinol in Patients with Hyperuricemia and Gout." *New England Journal of Medicine* 353, no. 23 (2005): 2450–61.

Begossi, A., N. Hanazaki, and R. M. Ramos. "Food Chain and the Reasons for Fish Food Taboos Among Amazonian and Atlantic Forest Fishers (Brazil)." *Ecological Applications* 14, no. 5 (2004): 1334–43.

"Being Gluten-Free 'Is Determined by Evolution,' Says Gastroenterologist." Accessed November 23, 2012. www.science20.com/news_articles/being_gluten free_determined_evolution_says_gastroenterologist-91578.

Bell, M. V., R. J. Henderson, and J. R. Sargent. "The Role of Polyunsaturated Fatty Acids in Fish." *Comparative Biochemistry and Physiology, Part B: Biochemistry and Molecular Biology* 83, no. 4 (1986): 711–19. doi:10.1016/0305-0491(86)90135-5.

Belluco, Simone, Carmen Losasso, Michela Maggioletti, Cristiana C. Alonzi, Maurizio G. Paoletti, and Antonia Ricci. "Edible Insects in a Food Safety and Nutritional Perspective: A Critical Review." *Comprehensive Reviews in Food Science and Food Safety* 12, no. 3 (2013): 296–313.

Bendini, Alessandra, Lorenzo Cerretani, Alegria Carrasco-Pancorbo, Ana Maria Gómez-Caravaca, Antonio Segura-Carretero, Alberto Fernández-Gutiérrez, and Giovanni Lercker. "Phenolic Molecules in Virgin Olive Oils: A Survey of Their Sensory Properties, Health Effects, Antioxidant Activity and Analytical Methods." *Molecules* 12, no. 8 (August 6, 2007): 1679–719. doi:10.3390/12081679.

Berdoy, M., J. P. Webster, and D. W. Macdonald. "Fatal Attraction in Rats Infected with Toxoplasma Gondii." *Proceedings of the Royal Society of London, Series B: Biological Sciences* 267, no. 1452 (August 7, 2000): 1591–94. doi:10.1098/rspb .2000.1182.

Bewley, J. Derek, Michael J. Black, and Peter Halmer. *The Encyclopedia of Seeds: Science, Technology and Uses.* CABI, 2006.

Bhasin, Shalender, Thomas W. Storer, Nancy Berman, Carlos Callegari, Brenda Clevenger, Jeffrey Phillips, Thomas J. Bunnell, Ray Tricker, Aida Shirazi, and Richard Casaburi. "The Effects of Supraphysiologic Doses of Testosterone on Muscle Size and Strength in Normal Men." *New England Journal of Medicine* 335, no. 1 (1996): 1–7. doi:10.1056/NEJM199607043350101.

Billing, J., and P. W. Sherman. "Antimicrobial Functions of Spices: Why Some Like It Hot." *Quarterly Review of Biology* 73, no. 1 (1998): 3–49.

Bishnoi, S., N. Khetarpaul, and R. K. Yadav. "Effect of Domestic Processing and Cooking Methods on Phytic Acid and Polyphenol Contents of Pea Cultivars (*Pisum sativum*)." *Plant Foods for Human Nutrition* 45, no. 4 (June 1, 1994): 381–88. doi:10.1007/BF01088088.

Björn, Lars Olof. "Vitamin D: Photobiological and Ecological Aspects." In *Photobiology*, edited by Lars Olof Björn, 531–52. Springer New York, 2008.

Blackley, Charles Harrison. *Experimental Researches on the Causes and Nature of Catarrhus Æstivus.* Ballière, Tindal & Cox, 1873. https://archive.org/details/experimentalres00blacgoog.

Blasbalg, Tanya L., Joseph R. Hibbeln, Christopher E. Ramsden, Sharon F. Majchrzak, and Robert R. Rawlings. "Changes in Consumption of Omega-3 and Omega-6 Fatty Acids in the United States During the 20th Century." *American Journal of Clinical Nutrition* 93, no. 5 (May 1, 2011): 950–62. doi:10.3945/ajcn.110.006643.

Blaser, Martin J. *Missing Microbes: How the Overuse of Antibiotics Is Fueling Our Modern Plagues.* Macmillan, 2014.

Bloch, Talia. "Could Kashrut Be Partly to Blame for Crohn's Disease?" *Jewish Daily Forward.* August 12, 2011. http://forward.com/articles/140645/could-kashrut-be-partly-to-blame-for-crohns-diseas/.

Bloomfield, S. F., R. Stanwell-Smith, R. W. R. Crevel, and J. Pickup. "Too Clean, or Not Too Clean: The Hygiene Hypothesis and Home Hygiene." *Clinical and Experimental Allergy* 36, no. 4 (2006): 402–25. doi:10.1111/j.1365-2222.2006.02463.x.

Blount, Roy, Jr. *Soupsongs/Webster's Ark.* Houghton Mifflin, 1987.

Bollet, A. J. "Politics and Pellagra: The Epidemic of Pellagra in the U.S. in the Early Twentieth Century." *Yale Journal of Biology and Medicine* 65, no. 3 (1992): 211–21.

Bovet, P., and F. Paccaud. "Commentary: Alcohol, Coronary Heart Disease and Public Health: Which Evidence-Based Policy." *International Journal of Epidemiology* 30, no. 4 (2001): 734–37.

Bower, Sylvia, Mary Kay Sharrett, and Steve Plogsted. *Celiac Disease: A Guide to Living with Gluten Intolerance.* Demos Medical Publishing, 2006.

Bramble, D. M., and D. E. Lieberman. "Endurance Running and the Evolution of *Homo.*" *Nature* 432, no. 7015 (2004): 345–52.

Bremer, Andrew A., Michele Mietus-Snyder, and Robert H. Lustig. "Toward a Unifying Hypothesis of Metabolic Syndrome." *Pediatrics* 129, no. 3 (March 1, 2012): 557–70. doi:10.1542/peds.2011-2912.

Brink, Pamela J. "The Fattening Room Among the Annang of Nigeria." *Medical Anthropology* 12, no. 1 (1989): 131–43. doi:10.1080/01459740.1989.9966016.

Brooks, Collin, Neil Pearce, and Jeroen Douwes. "The Hygiene Hypothesis in Allergy and Asthma: An Update." *Current Opinion in Allergy and Clinical Immunology* 13, no. 1 (2013): 70–77.

Brown, E. V. L. "Use-Abuse Theory of Changes in Refraction Versus Biologic Theory." *Archives of Ophthalmology* 28, no. 5 (1942): 845.

Bruyn, George William, and Charles M. Poser. *The History of Tropical Neurology: Nutritional Disorders.* Watson Publishing International, 2003.

Bryson, Peter D. *Comprehensive Reviews in Toxicology: For Emergency Clinicians.* CRC Press, 1996.

Buettner, D. *The Blue Zones: Lessons for Living Longer from the People Who've Lived the Longest.* National Geographic, 2010.

Burckhardt, John Lewis. *Notes on the Bedouins and Wahábys.* H. Colburn and R. Bentley, 1830.

Busfield, Frances, David L. Duffy, Janine B. Kesting, Shelley M. Walker, Paul K. Lovelock, David Good, Heather Tate, Denise Watego, Maureen Marczak, Noel Hayman, and Joanne T. E. Shaw. "A Genomewide Search for Type 2 Diabetes–Susceptibility Genes in Indigenous Australians." *American Journal of Human Genetics* 70, no. 2 (January 2, 2002): 349–57. doi:10.1086/338626.

Buxton, M. "Fish-Eating in Medieval England." In *Fish, Food from the Waters: Proceedings of the Oxford Symposium on Food and Cooking 1997,* edited by Harlan Walker, 51. Prospect Books, 1998.

Byremo, G., G. Rød, and K. H. Carlsen. "Effect of Climatic Change in Children with Atopic Eczema." *Allergy* 61, no. 12 (2006): 1403–10. doi:10.1111/j.1398-9995.2006.01209.x.

Calder, Philip C. "The Role of Marine Omega-3 (n-3) Fatty Acids in Inflammatory Processes, Atherosclerosis and Plaque Stability." *Molecular Nutrition and Food Research* 56, no. 7 (2012): 1073–80. doi:10.1002/mnfr.201100710.

Camargo Carlos A., Jr., Sunday Clark, Michael S. Kaplan, Philip Lieberman, and Robert A. Wood. "Regional Differences in EpiPen Prescriptions in the United

States: The Potential Role of Vitamin D." *Journal of Allergy and Clinical Immunology* 120, no. 1 ( July 2007): 131–36. doi:10.1016/j.jaci.2007.03.049.

Cardillo, M., and A. Lister. "Death in the Slow Lane." *Nature* 419, no. 6906 (2002): 440.

Carpenter, Kenneth John. *Beriberi, White Rice, and Vitamin B: A Disease, a Cause, and a Cure.* University of California Press, 2000.

Carrier, D. R., A. K. Kapoor, T. Kimura, M. K. Nickels, Satwanti, E. C. Scott, J. K. So, and E. Trinkaus. "The Energetic Paradox of Human Running and Hominid Evolution [and Comments and Reply]." *Current Anthropology* 25, no. 4 (1984): 483–95.

Casas-Agustench, Patricia, Albert Salas-Huetos, and Jordi Salas-Salvadó. "Mediterranean Nuts: Origins, Ancient Medicinal Benefits and Symbolism." *Public Health Nutrition* 14, no. 12A (2011): 2296–301.

Cassidy, Claire M. "The Good Body: When Big Is Better." *Medical Anthropology* 13, no. 3 (1991): 181–213. doi:10.1080/01459740.1991.9966048.

Catassi, Carlo, Julio C. Bai, Bruno Bonaz, Gerd Bouma, Antonio Calabrò, Antonio Carroccio, Gemma Castillejo, Carolina Ciacci, Fernanda Cristofori, Jernej Dolinsek, Ruggiero Francavilla, Luca Elli, Peter Green, Wolfgang Holtmeier, Peter Koehler, Sibylle Koletzko, Christof Meinhold, David Sanders, Michael Schumann, Detlef Schuppan, Reiner Ullrich, Andreas Vécsei, Umberto Volta, Victor Zevallos, Anna Sapone, and Alessio Fasano. "Non-Celiac Gluten Sensitivity: The New Frontier of Gluten Related Disorders." *Nutrients* 5, no. 10 (2013): 3839–53.

Catling, L., Ibrahim Abubakar, I. Lake, Louise Swift, and P. Hunter. "A Systematic Review of Analytical Observational Studies Investigating the Association Between Cardiovascular Disease and Drinking Water Hardness." *Journal of Water and Health* 6, no. 4 (2008): 433–42.

Cerqueira, Fernanda M., and Alicia J. Kowaltowski. "Mitochondrial Metabolism in Aging: Effect of Dietary Interventions." *Ageing Research Reviews* 12, no. 1 (2013): 22–28.

Cerqueira, Fernanda M., Fernanda M. da Cunha, Camille C. Caldeira da Silva, Bruno Chausse, Renato L. Romano, Camila Garcia, Pio Colepicolo, Marisa HG Medeiros, and Alicia J. Kowaltowski. "Long-Term Intermittent Feeding, but Not Caloric Restriction, Leads to Redox Imbalance, Insulin Receptor Nitration, and Glucose Intolerance." *Free Radical Biology and Medicine* 51, no. 7 (2011): 1454–60.

Cerulli, Ernesta. *Peoples of South-West Ethiopia and Its Borderland.* University Microfilms International, 1982.

Chaimanee, Yaowalak, Olivier Chavasseau, K. Christopher Beard, Aung Aung

Kyaw, Aung Naing Soe, Chit Sein, Vincent Lazzari, Laurent Marivaux, Bernard Marandat, Myat Swe, Mana Rugbumrung, Thit Lwin, Xavier Valentin, Zin-Maung-Maung-Thein, and Jean-Jacques Jaeger. "Late Middle Eocene Primate from Myanmar and the Initial Anthropoid Colonization of Africa." *Proceedings of the National Academy of Sciences* 109, no. 26 (June 26, 2012): 10293–97. doi:10.1073/pnas.1200644109.

Chakrabarti, S. D., R. Ganguly, S. K. Chatterjee, and A. Chakravarty. "Is Squatting a Triggering Factor for Stroke in Indians?" *Acta Neurologica Scandinavica* 105, no. 2 (2002): 124–27.

Chausse, Bruno, Carina Solon, Camille C. Caldeira da Silva, Ivan G. Masselli dos Reis, Fúlvia B. Manchado-Gobatto, Claudio A. Gobatto, Licio A. Velloso, and Alicia J. Kowaltowski. "Intermittent Fasting Induces Hypothalamic Modifications Resulting in Low Feeding Efficiency, Low Body Mass and Overeating." *Endocrinology* 155, no. 7 (July 5, 2014): 2456–66. http://press.endocrine.org/doi/abs/10.1210/en.2013-2057.

Chen, H., J.e.a. Common, R.l. Haines, A. Balakrishnan, S.j. Brown, C.s.m. Goh, H.j. Cordell, A. Sandilands, L. E. Campbell, K. Kroboth, A. D. Irvine, D. L. M. Goh, M. B. Y. Tang, H. P. van Bever, Y. C. Giam, W. H. I. McLean, and E. B. Lane. "Wide Spectrum of Filaggrin-Null Mutations in Atopic Dermatitis Highlights Differences Between Singaporean Chinese and European Populations." *British Journal of Dermatology* 165, no. 1 (2011): 106–14. doi:10.1111/j.1365-2133.2011.10331.x.

Chen, W., S. M. Dawsey, Y.-L. Qiao, S. D. Mark, Z.-W. Dong, P. R. Taylor, P. Zhao, and C. C. Abnet. "Prospective Study of Serum 25(OH)-Vitamin D Concentration and Risk of Oesophageal and Gastric Cancers." *British Journal of Cancer* 97, no. 1 (2007): 123–28. doi:10.1038/sj.bjc.6603834.

Cheney, Liana de Girolami. "The Oyster in Dutch Genre Paintings: Moral or Erotic Symbolism." *Artibus et Historiae* 8, no. 15 (1987): 135–58.

Choi, H. K., X. Gao, and G. Curhan. "Vitamin C Intake and the Risk of Gout in Men: A Prospective Study." *Archives of Internal Medicine* 169, no. 5 (March 9, 2009): 502–7. doi:10.1001/archinternmed.2008.606.

Chow, Ching Kuang. "Dietary Intake of Menaquinones and Risk of Cancer Incidence and Mortality." *American Journal of Clinical Nutrition* 92, no. 6 (December 1, 2010): 1533–34. doi:10.3945/ajcn.110.002337.

Cisterna, B., F. Flach, L. Vecchio, S. M. L. Barabino, S. Battistelli, T. E. Martin, M. Malatesta, and M. Biggiogera. "Can a Genetically Modified Organism–Containing Diet Influence Embryo Development? A Preliminary Study on Pre-Implantation Mouse Embryos." *European Journal of Histochemistry* 52, no. 4 (2009): 263–67.

Clarke, Charlotte Bringle. *Edible and Useful Plants of California.* University of California Press, 1977.

Clarke, Philip A. *Aboriginal Plant Collectors: Botanists and Australian Aboriginal People in the Nineteenth Century.* Rosenberg Publishing, 2008.

Cohen, M. N. "Introduction: Rethinking the Origins of Agriculture." *Current Anthropology* 50, no. 5 (2009): 591–95.

Cooper, P. J. "Interactions between Helminth Parasites and Allergy." *Current Opinion in Allergy and Clinical Immunology* 9, no. 1 (2009): 29.

————. "Intestinal Worms and Human Allergy." *Parasite Immunology* 26, no. 11–12 (2004): 455–67. doi:10.1111/j.0141-9838.2004.00728.x.

Correale, Jorge, and Mauricio Farez. "Association Between Parasite Infection and Immune Responses in Multiple Sclerosis." *Annals of Neurology* 61, no. 2 (2007): 97–108. doi:10.1002/ana.21067.

Cui, Jie, Xinpu Yuan, Lina Wang, Gareth Jones, and Shuyi Zhang. "Recent Loss of Vitamin C Biosynthesis Ability in Bats." *PLoS ONE* 6, no. 11 (November 1, 2011): e27114. doi:10.1371/journal.pone.0027114.

Cui, Jie, Yi-Hsuan Pan, Yijian Zhang, Gareth Jones, and Shuyi Zhang. "Progressive Pseudogenization: Vitamin C Synthesis and Its Loss in Bats." *Molecular Biology and Evolution* 28, no. 2 (February 1, 2011): 1025–31. doi:10.1093/molbev/msq286.

Cunningham, C. B., N. Schilling, C. Anders, and D. R. Carrier. "The Influence of Foot Posture on the Cost of Transport in Humans." *Journal of Experimental Biology* 213, no. 5 (March 1, 2010): 790–97. doi:10.1242/jeb.038984.

Curtis, Robert I. "Umami and the Foods of Classical Antiquity." *American Journal of Clinical Nutrition* 90, no. 3 (September 1, 2009): 712S–18S. doi:10.3945/ajcn.2009.27462C.

Curtis, Val, Robert Aunger, and Tamer Rabie. "Evidence That Disgust Evolved to Protect from Risk of Disease." *Proceedings of the Royal Society of London, Series B: Biological Sciences* 271, Supp. 4 (2004): S131–33.

Davenport, John. *Aphrodisiacs and Anti-Aphrodisiacs: Three Essays on the Powers of Reproduction; with Some Account of the Judicial "Congress" as Practiced in France During the Seventeenth Century.* Privately printed, 1869.

Davidson, Daniel Sutherland. "Australian Throwing-Sticks, Throwing-Clubs, and Boomerangs." *American Anthropologist* 38, no. 1 (1936): 76–100.

De Garine, Igor, and Georgius J. A. Koppert. "*Guru*-Fattening Sessions Among the Massa." *Ecology of Food and Nutrition* 25, no. 1 (1991): 1–28. doi:10.1080/03670244.1991.9991151.

De Graaf, L., A. H. P. M. Brouwers, and W. L. Diemont. "Is Decreased Libido Associated with the Use of HMG-CoA-Reductase Inhibitors?" *British Journal of Clinical Pharmacology* 58, no. 3 (2004): 326–28.

De Gruijl, F. R. "Skin Cancer and Solar UV Radiation." *European Journal of Cancer* 35, no. 14 (December 1999): 2003–9. doi:10.1016/S0959-8049(99)00283-X.

De Oliveira, Erick Prado, and Roberto Carlos Burini. "The Impact of Physical Exercise on the Gastrointestinal Tract." *Current Opinion in Clinical Nutrition and Metabolic Care* 12, no. 5 (2009): 533–38.

De Vendômois, Joël Spiroux, François Roullier, Dominique Cellier, and Gilles-Eric Séralini. "A Comparison of the Effects of Three GM Corn Varieties on Mammalian Health." *International Journal of Biological Sciences* 5, no. 7 (2009): 706.

Decker, Evalotte, Guido Engelmann, Annette Findeisen, Patrick Gerner, Martin Laaß, Dietrich Ney, Carsten Posovszky, Ludwig Hoy, and Mathias W. Hornef. "Cesarean Delivery Is Associated with Celiac Disease but Not Inflammatory Bowel Disease in Children." *Pediatrics* 125, no. 6 (June 1, 2010): e1433–40. doi:10.1542/peds.2009-2260.

Delany, Paul. "Constantinus Africanus' *De Coitu*: A Translation." *Chaucer Review* 4, no. 1 (Summer 1969): 55–65.

DeLong, Gayle. "Conflicts of Interest in Vaccine Safety Research." *Accountability in Research* 19, no. 2 (2012): 65–88.

Dennis, J. V. "If You Drink, Don't Fly: Fermented Fruit and Sap Can Inebriate Birds." *Birder's World* 1 (1987): 15–19.

Deshpande, S. S. *Handbook of Food Toxicology*. CRC Press, 2002.

Devereux, Graham, Augusto A. Litonjua, Stephen W. Turner, Leone C. A. Craig, Geraldine McNeill, Sheelagh Martindale, Peter J. Helms, Anthony Seaton, and Scott T. Weiss. "Maternal Vitamin D Intake During Pregnancy and Early Childhood Wheezing." *American Journal of Clinical Nutrition* 85, no. 3 (March 2007): 853–59.

Diamond, Jared M. "Archaeology: Talk of Cannibalism." *Nature* 407, no. 6800 (2000): 25–26.

———. *Collapse: How Societies Choose to Fail or Succeed*. Paw Prints, 2008.

———. *Guns, Germs, and Steel: The Fates of Human Societies*. W. W. Norton, 1997.

———. "Palaeontology: The Last Giant Kangaroo." *Nature* 454, no. 7206 (August 13, 2008): 835–36. doi:10.1038/454835a.

Dickson, Harold Richard Patrick. *The Arab of the Desert: A Glimpse into Badawin Life in Kuwait and Sau'di Arabia*. Allen & Unwin, 1959.

Diels, Johan, Mario Cunha, Célia Manaia, Bernardo Sabugosa-Madeira, and Margarida Silva. "Association of Financial or Professional Conflict of Interest to Research Outcomes on Health Risks or Nutritional Assessment Studies of Genetically Modified Products." *Food Policy* 36, no. 2 (April 2011): 197–203. doi:10.1016/j.foodpol.2010.11.016.

DiMarino, Anthony J., Jr., and Stanley B. Benjamin, eds. *Gastrointestinal Disease: An Endoscopic Approach.* SLACK, 2002.

Dirani, M., L. Tong, G. Gazzard, X. Zhang, A. Chia, T. L. Young, K. A. Rose, P. Mitchell, and S.-M. Saw. "Outdoor Activity and Myopia in Singapore Teenage Children." *British Journal of Ophthalmology* 93, no. 8 (August 1, 2009): 997–1000. doi:10.1136/bjo.2008.150979.

Dixon, J. B., G. J. Egger, E. A. Finkelstein, J. G. Kral, and G. W. Lambert. "'Obesity Paradox' Misunderstands the Biology of Optimal Weight Throughout the Life Cycle." *International Journal of Obesity* 39 (2015): 82–84. doi:10.1038/ijo.2014.59.

Dobney, K., and A. Ervynck. "To Fish or Not to Fish? Evidence for the Possible Avoidance of Fish Consumption During the Iron Age around the North Sea." In *The Later Iron Age in Britain and Beyond*, edited by C. Haselgrove and T. Moore, 403–18. Oxbow Books, 2007. www.vliz.be/imis/imis.php?module=ref&refid=110515.

Domingo, José L., and Jordi Giné Bordonaba. "A Literature Review on the Safety Assessment of Genetically Modified Plants." *Environment International* 37, no. 4 (May 2011): 734–42. doi:10.1016/j.envint.2011.01.003.

Drewnowski, Adam, and Carmen Gomez-Carneros. "Bitter Taste, Phytonutrients, and the Consumer: A Review." *American Journal of Clinical Nutrition* 72, no. 6 (December 1, 2000): 1424–35.

Drouin, Guy, Jean-Rémi Godin, and Benoit Pagé. "The Genetics of Vitamin C Loss in Vertebrates." *Current Genomics* 12, no. 5 (August 1, 2011): 371–78. doi:10.2174/138920211796429736.

Drummond, Jack C., and Anne Wilbraham. *The Englishman's Food: A History of Five Centuries of English Diet.* Pimlico, 1991.

Duboucher, C., R. Escamilla, F. Rocchiccioli, A. Negre, A. Lageron, and J. Migueres. "Pulmonary Lipogranulomatosis Due to Excessive Consumption of Apples." *CHEST Journal* 90, no. 4 (1986): 611–12.

Dugas, Lara R., Regina Harders, Sarah Merrill, Kara Ebersole, David A. Shoham, Elaine C. Rush, Felix K. Assah, Terrence Forrester, Ramon A. Durazo-Arvizu, and Amy Luke. "Energy Expenditure in Adults Living in Developing Compared with Industrialized Countries: A Meta-Analysis of Doubly Labeled Water Studies." *American Journal of Clinical Nutrition* 93, no. 2 (February 1, 2011): 427–41. doi:10.3945/ajcn.110.007278.

Duhan, Arti, Bhag Mal Chauhan, Darshan Punia, and Amin Chand Kapoor. "Phytic Acid Content of Chickpea (*Cicer arietinum*) and Black Gram (*Vigna mungo*): Varietal Differences and Effect of Domestic Processing and Cooking Methods."

*Journal of the Science of Food and Agriculture* 49, no. 4 (January 1, 1989): 449–55. doi:10.1002/jsfa.2740490407.

Dunstan, Janet A., Trevor A. Mori, Anne Barden, Lawrence J. Beilin, Angie L. Taylor, Patrick G. Holt, and Susan L. Prescott. "Fish Oil Supplementation in Pregnancy Modifies Neonatal Allergen-Specific Immune Responses and Clinical Outcomes in Infants at High Risk of Atopy." *Journal of Allergy and Clinical Immunology* 112, no. 6 (December 2003): 1178–84. doi:10.1016/j.jaci.2003.09.009.

DuPuis, E. Melanie. *Nature's Perfect Food: How Milk Became America's Drink.* New York University Press, 2002.

Eaton, S. B., S. B. Eaton, A. J. Sinclair, L. Cordain, and N. J. Mann. "Dietary Intake of Long-Chain Polyunsaturated Fatty Acids During the Paleolithic." *World Review of Nutrition and Dietetics* 83 (1998): 12–23.

Eizirik, E., W. J. Murphy, and S. J. O'Brien. "Molecular Dating and Biogeography of the Early Placental Mammal Radiation." *Journal of Heredity* 92, no. 2 (2001): 212–19.

Elgar, Mark A., and Bernard J. Crespi. *Cannibalism: Ecology and Evolution Among Diverse Taxa.* Oxford University Press, 1992.

Elliott, Charlene. "Canada's Great Butter Caper: On Law, Fakes and the Biography of Margarine." *Food, Culture and Society: An International Journal of Multidisciplinary Research* 12, no. 3 (2009): 379–96.

Elmore, J. G., and A. R. Feinstein. "Joseph Goldberger: An Unsung Hero of American Clinical Epidemiology." *Annals of Internal Medicine* 121, no. 5 (1994): 372–75.

Elwood, J. Mark, and Janet Jopson. "Melanoma and Sun Exposure: An Overview of Published Studies." *International Journal of Cancer* 73, no. 2 (1997): 198–203. doi:10.1002/(SICI)1097-0215(19971009)73:2<198::AID-IJC6>3.0.CO;2-R.

Emsley, John. *Molecules of Murder: Criminal Molecules and Classic Cases.* Royal Society of Chemistry, 2008.

Eppinger, Michael. *Field Guide to Wild Flowers of Britain and Europe.* New Holland Publishers, 2007.

Erkkola, M., M. Kaila, B. I. Nwaru, C. Kronberg-Kippilä, S. Ahonen, J. Nevalainen, R. Veijola, J. Pekkanen, J. Ilonen, O. Simmel, M. Knip, and S. M. Virtanen. "Maternal Vitamin D Intake During Pregnancy Is Inversely Associated with Asthma and Allergic Rhinitis in 5-Year-Old Children." *Clinical and Experimental Allergy* 39, no. 6 (2009): 875–82. doi:10.1111/j.1365-2222.2009.03234.x.

"Ethanol/Corn Balance Sheets—Agricultural Marketing Resource Center." Accessed August 11, 2014. www.agmrc.org/renewable_energy/ethanol/ethanol-corn-balance-sheets/.

Fairclough, Stephen H., and Kim Houston. "A Metabolic Measure of Mental Effort." *Biological Psychology* 66, no. 2 (2004): 177–90.

Falchi, Mario, Julia Sarah El-Sayed Moustafa, Petros Takousis, Francesco Pesce, Amélie Bonnefond, Johanna C. Andersson-Assarsson, Peter H. Sudmant, et al. "Low Copy Number of the Salivary Amylase Gene Predisposes to Obesity." *Nature Genetics* 46, no. 5 (2014): 492–97.

Falk, Dean, John C. Redmond Jr, John Guyer, C. Conroy, Wolfgang Recheis, Gerhard W. Weber, and Horst Seidler. "Early Hominid Brain Evolution: A New Look at Old Endocasts." *Journal of Human Evolution* 38, no. 5 (2000): 695–717.

Fallon, Sally, and Mary G. Enig. *Nourishing Traditions: The Cookbook That Challenges Politically Correct Nutrition and the Diet Dictocrats.* 2nd ed. NewTrends, 1999.

"FastStats: Body Measurements." Accessed August 26, 2014. www.cdc.gov/nchs /fastats/body-measurements.htm.

Faye, B. "The Sustainability Challenge to the Dairy Sector—The Growing Importance of Non-Cattle Milk Production Worldwide." *International Dairy Journal* 24, no. 2 (June 1, 2012): 50–56.

Fediuk, Karen, Nick Hidiroglou, René Madère, and Harriet V. Kuhnlein. "Vitamin C in Inuit Traditional Food and Women's Diets." *Journal of Food Composition and Analysis* 15, no. 3 (June 2002): 221–35. doi:10.1006/jfca.2002.1053.

Fenwick, Dorothy E., and David Oakenfull. "Saponin Content of Food Plants and Some Prepared Foods." *Journal of the Science of Food and Agriculture* 34, no. 2 (1983): 186–91. doi:10.1002/jsfa.2740340212.

Fernando, Ranjan J., Sujatha S. E. Fernando, and Anthony S.-Y. Leong. *Tropical Infectious Diseases: Epidemiology, Investigation, Diagnosis and Management.* Cambridge University Press, 2001.

Fernando, Shanaka, and Greg Ronald Hill. *Lentil as Anything: Everybody Deserves a Place at the Table.* Vivid Publishing, 2012.

Fessler, Daniel, and Kevin Haley. "Guarding the Perimeter: The Outside-inside Dichotomy in Disgust and Bodily Experience." *Cognition and Emotion* 20, no. 1 (2006): 3–19. doi:10.1080/02699930500215181.

Feynman, Joan, and Alexander Ruzmaikin. "Climate Stability and the Development of Agricultural Societies." *Climatic Change* 84, no. 3–4 (2007): 295–311.

Figueiredo, Camila Alexandrina, Mauricio L. Barreto, Laura C. Rodrigues, Philip J. Cooper, Nívea Bispo Silva, Leila D. Amorim, and Neuza Maria Alcantara-Neves. "Chronic Intestinal Helminth Infections Are Associated with Immune Hyporesponsiveness and Induction of a Regulatory Network." *Infection and Immunity* 78, no. 7 (July 1, 2010): 3160–67. doi:10.1128/IAI.01228-09.

Flegal K. M., B. K. Kit, H. Orpana, and B. I. Graubard. "Association of All-Cause Mortality with Overweight and Obesity Using Standard Body Mass Index Categories: A Systematic Review and Meta-Analysis." *JAMA* 309, no. 1 (January 2, 2013): 71–82. doi:10.1001/jama.2012.113905.

Fontana, Luigi. "Long-Term Effects of Calorie or Protein Restriction on Serum IGF-1 and IGFBP-3 Concentration in Humans." *Aging Cell* 7, no. 5 (October 1, 2008): 681–87.

Fox, Margalit. "Frank C. Garland, 60, Who Connected Vitamin D Deficiency and Cancer, Dies." *New York Times*, September 4, 2010. www.nytimes.com/2010/09/05/us/05garland.html.

Frederick, David A., Daniel M. T. Fessler, and Martie G. Haselton. "Do Representations of Male Muscularity Differ in Men's and Women's Magazines?" *Body Image* 2, no. 1 (2005): 81–86.

Freedman, Neal D., Yikyung Park, Christian C. Abnet, Albert R. Hollenbeck, and Rashmi Sinha. "Association of Coffee Drinking with Total and Cause-Specific Mortality." *New England Journal of Medicine* 366, no. 20 (2012): 1891–904. doi:10.1056/NEJMoa1112010.

French, Jackie. *The Camel Who Crossed Australia*. HarperCollins Australia, 2010.

Fried, Susan K., and Salome P. Rao. "Sugars, Hypertriglyceridemia, and Cardiovascular Disease." *American Journal of Clinical Nutrition* 78, no. 4 (2003): 873S–880S.

Frisch, Rose E. *Female Fertility and the Body Fat Connection*. University of Chicago Press, 2004.

Fryxell, J. M., and A. R. E. Sinclair. "Causes and Consequences of Migration by Large Herbivores." *Trends in Ecology and Evolution* 3, no. 9 (September 1988): 237–41. doi:10.1016/0169-5347(88)90166-8.

Fuemmeler, Bernard F., Margaret K. Pendzich, and Kenneth P. Tercyak. "Weight, Dietary Behavior, and Physical Activity in Childhood and Adolescence: Implications for Adult Cancer Risk." *Obesity Facts* 2, no. 3 (2009): 179–86. doi:10.1159/000220605.

Fujiwara, Miyuki, Satoshi Hasebe, Risa Nakanishi, Kohhei Tanigawa, and Hiroshi Ohtsuki. "Seasonal Variation in Myopia Progression and Axial Elongation: An Evaluation of Japanese Children Participating in a Myopia Control Trial." *Japanese Journal of Ophthalmology* 56, no. 4 (July 1, 2012): 401–6. doi:10.1007/s10384-012-0148-1.

Fulgoni, Victor L. "Current Protein Intake in America: Analysis of the National Health and Nutrition Examination Survey, 2003–2004." *American Journal of Clinical Nutrition* 87, no. 5 (2008): 1554S–1557S.

Fuller, Thomas C., and Elizabeth May McClintock. *Poisonous Plants of California (California Natural History Guides)*. University of California Press, 1986.

Gade, D. W. "Llamas and Alpacas." *The Cambridge World History of Food and Nutrition*, 2000, 555–59.

———. *Nature and Culture in the Andes*. University of Wisconsin Press, 1999.

Gaillard, Yvan, and Gilbert Pepin. "Poisoning by Plant Material: Review of Human

Cases and Analytical Determination of Main Toxins by High-Performance Liquid Chromatography–(tandem) Mass Spectrometry." *Journal of Chromatography B: Biomedical Sciences and Applications* 733, nos. 1–2 (October 15, 1999): 181–229. doi:10.1016/S0378-4347(99)00181-4.

Gale, C. R., S. M. Robinson, N. C. Harvey, M. K. Javaid, B. Jiang, C. N. Martyn, K. M. Godfrey, and C. Cooper. "Maternal Vitamin D Status During Pregnancy and Child Outcomes." *European Journal of Clinical Nutrition* 62, no. 1 (2008): 68–77. doi:10.1038/sj.ejcn.1602680.

Gandini, Sara, Francesco Sera, Maria Sofia Cattaruzza, Paolo Pasquini, Orietta Picconi, Peter Boyle, and Carmelo Francesco Melchi. "Meta-Analysis of Risk Factors for Cutaneous Melanoma: II. Sun Exposure." *European Journal of Cancer* 41, no. 1 (January 2005): 45–60. doi:10.1016/j.ejca.2004.10.016.

Gibbons, Whit, Robert R. Haynes, and Joab L. Thomas. *Poisonous Plants and Venomous Animals of Alabama and Adjoining States.* University of Alabama Press, 1990.

Gibson, Peter R., and Susan J. Shepherd. "Food Choice as a Key Management Strategy for Functional Gastrointestinal Symptoms." *American Journal of Gastroenterology* 107, no. 5 (2012): 657–66.

Gire, Stephen K., Augustine Goba, Kristian G. Andersen, Rachel SG Sealfon, Daniel J. Park, Lansana Kanneh, Simbirie Jalloh, et al. "Genomic Surveillance Elucidates Ebola Virus Origin and Transmission During the 2014 Outbreak." *Science* 345, no. 6202 (2014): 1369–72.

"Glossary of Sexual and Scatological Euphemisms." Accessed November 30, 2012. www.uta.fi/FAST/GC/sex-scat.html.

Goldberger, J., and G. A. Wheeler. *The Experimental Production of Pellagra in Human Subjects by Means of Diet.* U.S. Public Health Service Hygienic Laboratory Bulletin no. 120, February 1920.

Goldbohm, R. Alexandra, Astrid M. J. Chorus, Francisca Galindo Garre, Leo J. Schouten, and Piet A. van den Brandt. "Dairy Consumption and 10-Y Total and Cardiovascular Mortality: A Prospective Cohort Study in the Netherlands." *American Journal of Clinical Nutrition* 93, no. 3 (March 1, 2011): 615–27. doi:10.3945/ajcn.110.000430.

Gowlett, John A. J. "Special Issue: Innovation and the Evolution of Human Behavior. The Vital Sense of Proportion: Transformation, Golden Section, and 1: 2 Preference in Acheulean Bifaces." *PaleoAnthropology* 174 (2011): 187.

Grady, D., D. Herrington, V. Bittner, R. Blumenthal, M. Davidson, M. Hlatky, J. Hsia, S. Hulley, A. Herd, S. Khan, L. K. Newby, D. Waters, E. Vittinghoff, and N. Wenger, for the HERS Research Group. "Cardiovascular Disease Outcomes During 6.8 Years of Hormone Therapy: Heart and Estrogen/Progestin

Replacement Study Follow-up (HERS II)." *JAMA* 288, no. 1 (July 3, 2002): 49–57. doi:10.1001/jama.288.1.49.

Grady, D., S. M. Rubin, D. B. Petitti, C. S. Fox, D. Black, B. Ettinger, V. L. Ernster, and S. R. Cummings. "Hormone Therapy to Prevent Disease and Prolong Life in Postmenopausal Women." *Annals of Internal Medicine* 117, no. 12 (1992): 1016–37.

Grant, William B., and Connie M. Soles. "Epidemiologic Evidence for Supporting the Role of Maternal Vitamin D Deficiency as a Risk Factor for the Development of Infantile Autism." *Dermato-Endocrinology* 1, no. 4 (July 1, 2009): 223–28. doi:10.4161/derm.1.4.9500.

Graudal, Niels, Gesche Jürgens, Bo Baslund, and Michael H. Alderman. "Compared With Usual Sodium Intake, Low- and Excessive-Sodium Diets Are Associated With Increased Mortality: A Meta-Analysis." *American Journal of Hypertension* 27, no. 9 (September 1, 2014): 1129–37. doi:10.1093/ajh/hpu028.

Green, Adèle C., Gail M. Williams, Valerie Logan, and Geoffrey M. Strutton. "Reduced Melanoma After Regular Sunscreen Use: Randomized Trial Follow-up." *Journal of Clinical Oncology* 29, no. 3 (January 20, 2011): 257–63. doi:10.1200/JCO.2010.28.7078.

Gregory, Denis. *Australia's Great Explorers: Tales of Tragedy and Triumph.* Exisle Publishing, 2007.

Greim, Helmut, and Robert Snyder. *Toxicology and Risk Assessment: A Comprehensive Introduction.* John Wiley & Sons, 2008.

Grinin, Leonid Efimovich, Victor C. De Munck, and A. V. Korotaev. *History and Mathematics: Analyzing and Modeling Global Development.* Editorial URSS, 2006.

Grivas, T. B., E. Vasiliadis, V. Mouzakis, C. Mihas, and G. Koufopoulos. "Association Between Adolescent Idiopathic Scoliosis Prevalence and Age at Menarche in Different Geographic Latitudes." *Scoliosis* 1, no. 9 (2006). www.biomed central.com/content/pdf/1748-7161-1-9.pdf.

Grøntved, Anders, and Frank B. Hu. "Television Viewing and Risk of Type 2 Diabetes, Cardiovascular Disease, and All-Cause Mortality: A Meta-Analysis." *JAMA* 305, no. 23 (2011): 2448–55.

Guallar, Eliseo, JoAnn E. Manson, Christine Laine, and Cynthia Mulrow. "Postmenopausal Hormone Therapy: The Heart of the Matter." *Annals of Internal Medicine* 158, no. 1 (January 1, 2013): 69–70. doi:10.7326/0003-4819-158-1-2013 01010-00015.

Guallar, Eliseo, Saverio Stranges, Cynthia Mulrow, Lawrence J. Appel, and Edgar R. Miller III. "Enough Is Enough: Stop Wasting Money on Vitamin and Mineral Supplements." *Annals of Internal Medicine* 159, no. 12 (December 17, 2013): 850–51. doi:10.7326/0003-4819-159-12-201312170-00011.

Gunnell, D., S. E. Oliver, T. J. Peters, J. L. Donovan, R. Persad, M. Maynard, D. Gillatt, et al. "Are Diet–Prostate Cancer Associations Mediated by the IGF Axis? A Cross-Sectional Analysis of Diet, IGF-1 and IGFBP-3 in Healthy Middle-Aged Men." *British Journal of Cancer* 88, no. 11 (2003): 1682–86. doi:10.1038/sj.bjc.6600946.

Gupta, Atul, Andrew Bush, Catherine Hawrylowicz, and Sejal Saglani. "Vitamin D and Asthma in Children." *Paediatric Respiratory Reviews* 13, no. 4 (December 2012): 236–43. doi:10.1016/j.prrv.2011.07.003.

Gutiérrez, Yezid. *Diagnostic Pathology of Parasitic Infections: With Clinical Correlations.* 2nd ed. Oxford University Press, 2000.

Haboubi, Nadim. "Coeliac Disease: From A–Z." *Expert Opinon on Therapeutic Patients* 17, no. 7 (July 2007): 799–817.

Hallenbeck, Terri. "Vermont Defends GMO Labeling Law." August 8, 2014. www.burlingtonfreepress.com/story/news/politics/2014/08/08/gmo-lawsuit-response/13800873/.

Haque, M. R., and J. Howard Bradbury. "Total Cyanide Determination of Plants and Foods Using the Picrate and Acid Hydrolysis Methods." *Food Chemistry* 77, no. 1 (May 2002): 107–14. doi:10.1016/S0308-8146(01)00313-2.

Harari, Marco, Jashovam Shani, Vladimir Seidl, and Eugenia Hristakieva. "Climatotherapy of Atopic Dermatitis at the Dead Sea: Demographic Evaluation and Cost-Effectiveness." *International Journal of Dermatology* 39, no. 1 (2000): 59–69. doi:10.1046/j.1365-4362.2000.00840.x.

Hardus, Madeleine E., Adriano R. Lameira, Astri Zulfa, S. Suci Utami Atmoko, Han de Vries, and Serge A. Wich. "Behavioral, Ecological, and Evolutionary Aspects of Meat-Eating by Sumatran Orangutans (*Pongo abelii*)." *International Journal of Primatology* 33, no. 2 (2012): 287–304.

Harvey, Janice, and Inka Milewski. *Salmon Aquaculture in the Bay of Fundy: An Unsustainable Industry.* Conservation Council of New Brunswick, 2007. www.conservationcouncil.ca/publications/.

Hata, Tissa R., Paul Kotol, Michelle Jackson, Meggie Nguyen, Aimee Paik, Don Udall, Kimi Kanada, Kenshi Yamasaki, Doru Alexandrescu, and Richard L. Gallo. "Administration of Oral Vitamin D Induces Cathelicidin Production in Atopic Individuals." *Journal of Allergy and Clinical Immunology* 122, no. 4 (October 2008): 829–31. doi:10.1016/j.jaci.2008.08.020.

Hatzis, Christos M., Christopher Papandreou, Evridiki Patelarou, Constantine I. Vardavas, Eleni Kimioni, Dimitra Sifaki-Pistolla, Anna Vergetaki, and Anthony G. Kafatos. "A 50-Year Follow-up of the Seven Countries Study: Prevalence of Cardiovascular Risk Factors, Food and Nutrient Intakes Among Cretans." *Hormones* 12, no. 3 (September 2013): 379–85.

Haupt, Angela. "Ashton Kutcher's Fruitarian Diet: What Went Wrong?" *US News & World Report.* February 7, 2013. http://health.usnews.com/health-news/articles /2013/02/07/ashton-kutchers-fruitarian-diet-what-went-wrong.

Hawkes, Ernest William. *The Labrador Eskimo.* Canada Department of Mines, Geological Survey, 1916.

Hawkes, K., J. F. O'Connell, N. G. Blurton Jones, H. Alvarez, and E. L. Charnov. "Grandmothering, Menopause, and the Evolution of Human Life Histories." *Proceedings of the National Academy of Sciences* 95, no. 3 (February 3, 1998): 1336–39.

He, Ka, Shufa Du, Pengcheng Xun, Sangita Sharma, Huijun Wang, Fengying Zhai, and Barry Popkin. "Consumption of Monosodium Glutamate in Relation to Incidence of Overweight in Chinese Adults: China Health and Nutrition Survey (CHNS)." *American Journal of Clinical Nutrition* 93, no. 6 (June 1, 2011): 1328–36. doi:10.3945/ajcn.110.008870.

Hegsted, D. Mark. "Fractures, Calcium, and the Modern Diet." *American Journal of Clinical Nutrition* 74, no. 5 (November 1, 2001): 571–73.

Hehemann, Jan-Hendrik, Gaëlle Correc, Tristan Barbeyron, William Helbert, Mirjam Czjzek, and Gurvan Michel. "Transfer of Carbohydrate-Active Enzymes from Marine Bacteria to Japanese Gut Microbiota." *Nature* 464, no. 7290 (April 8, 2010): 908–12. doi:10.1038/nature08937.

Heizer, Robert Fleming, and Albert B. Elsasser. *The Natural World of the California Indians.* University of California Press, 1980.

Henrich, Joseph, and Natalie Henrich. "The Evolution of Cultural Adaptations: Fijian Food Taboos Protect against Dangerous Marine Toxins." *Proceedings of the Royal Society B: Biological Sciences* 277, no. 1701 (December 22, 2010): 3715–24. doi:10.1098/rspb.2010.1191.

Heyer, E., L. Brazier, L. Ségurel, T. Hegay, F. Austerlitz, L. Quintana-Murci, M. Georges, P. Pasquet, and M. Veuille. "Lactase Persistence in Central Asia: Phenotype, Genotype, and Evolution." *Human Biology* 83, no. 3 (2011): 379–92.

Higgins, Peter D. R. "Epidemiology of Constipation in North America: A Systematic Review." *American Journal of Gastroenterology* 99, no. 4 (2004): 750–59.

Hoberg, Eric P. "Phylogeny of *Taenia*: Species Definitions and Origins of Human Parasites." *Parasitology International* 55, Supplement (2006): S23–30. doi:10.1016/j .parint.2005.11.049.

Hofreiter, Michael, Eva Kreuz, Jonas Eriksson, Grit Schubert, and Gottfried Hohmann. "Vertebrate DNA in Fecal Samples from Bonobos and Gorillas: Evidence for Meat Consumption or Artefact?" *PLoS ONE* 5, no. 2 (February 25, 2010): e9419. doi:10.1371/journal.pone.0009419.

Holdaway, R. N., and C. Jacomb. "Rapid Extinction of the Moas (Aves: Dinornithiformes): Model, Test, and Implications." *Science* 287, no. 5461 (2000): 2250–54.

Holick, M. F. "Environmental Factors That Influence the Cutaneous Production of Vitamin D." *American Journal of Clinical Nutrition* 61, no. 3 (March 1, 1995): 638S–45S.

Holick, Michael F. "Vitamin D Deficiency." *New England Journal of Medicine* 357, no. 3 (July 19, 2007): 266–81. doi: 10.1056/NEJMra070553.

Hopkins, Jerry, Anthony Bourdain, and Michael A. Freeman. *Extreme Cuisine: The Weird and Wonderful Foods That People Eat.* Tuttle Publishing, 2004.

Hopkins, P. N. "Effects of Dietary Cholesterol on Serum Cholesterol: A Meta-Analysis and Review." *American Journal of Clinical Nutrition* 55, no. 6 (June 1, 1992): 1060–70.

Howard, Brian Clark. "Salmon Farming Gets Leaner and Greener." March 19, 2014. http://news.nationalgeographic.com/news/2014/03/140319-salmon-farming-sustainable-aquaculture/.

Hu, Frank B. "The Mediterranean Diet and Mortality—Olive Oil and Beyond." *New England Journal of Medicine* 348, no. 26 (June 26, 2003): 2595–96. doi:10.1056/NEJMp030069.

Hu, Frank B., Tricia Y. Li, Graham A. Colditz, Walter C. Willett, and JoAnn E. Manson. "Television Watching and Other Sedentary Behaviors in Relation to Risk of Obesity and Type 2 Diabetes Mellitus in Women." *JAMA* 289, no. 14 (2003): 1785–91.

Huang, Angela L., Xiaoke Chen, Mark A. Hoon, Jayaram Chandrashekar, Wei Guo, Dimitri Tränkner, Nicholas J. P. Ryba, and Charles S. Zuker. "The Cells and Logic for Mammalian Sour Taste Detection." *Nature* 442, no. 7105 (August 24, 2006): 934–38. doi:10.1038/nature05084.

Hulme, Frederick Edward. *Bards and Blossoms; or, The Poetry, History, and Associations of Flowers.* Marcus Ward, 1877.

Hypponen, E. "Infant Vitamin D Supplementation and Allergic Conditions in Adulthood: Northern Finland Birth Cohort 1966." *Annals of the New York Academy of Sciences* 1037 (2004): 84–95.

ICDDR. *Diarrhoeal Diseases Research.* International Centre for Diarrhoeal Diseases Research, Bangladesh, 1992.

IICA, CARDI, and MINAG. *Seminar on Research and Development of Fruit Trees (Citrus Excluded).* IICA Biblioteca Venezuela, June 1980.

Inoue, Masamichi S. *Okinawa and the U.S. Military: Identity Making in the Age of Globalization.* Columbia University Press, 2007.

Insawang, Tonkla, Carlo Selmi, Ubon Cha'on, Supattra Pethlert, Puangrat Yongvanit, Premjai Areejitranusorn, Patcharee Boonsiri, Tueanjit Khampitak,

Roongpet Tangrassameeprasert, Chadamas Pinitsoontorn, Vitoon Prasong-wattana, M. Eric Gershwin, and Bruce D. Hammock. "Monosodium Gluta-mate (MSG) Intake Is Associated with the Prevalence of Metabolic Syndrome in a Rural Thai Population." *Nutrition and Metabolism* 9, no. 1 (2012): 50.

Isaac, Barbara. "Throwing and Human Evolution." *African Archaeological Review* 5, no. 1 (1987): 3–17.

Iwu, Maurice M. *Handbook of African Medicinal Plants*. CRC Press, 1993.

Izhaki, Ido, and Uriel N. Safriel. "Why Are There So Few Exclusively Frugivo-rous Birds? Experiments on Fruit Digestibility." *Oikos* 54 (1989): 23–32.

Jacobi, Lora, and Thomas F. Cash. "In Pursuit of the Perfect Appearance: Dis-crepancies Among Self-Ideal Percepts of Multiple Physical Attributes." *Journal of Applied Social Psychology* 24, no. 5 (1994): 379–96.

Jacobs, Alan H. *The Pastoral Masai of Kenya*. University of Illinois Department of Anthropology, 1969.

"Japan Bluefin Tuna Fetches Record $1.7m." BBC, Asia section, January 5, 2013. www.bbc.co.uk/news/world-asia-20919306.

Javanbakht, Mohammad Hassan, Seyed Ali Keshavarz, Mahmoud Djalali, Fereydoun Siassi, Mohammad Reza Eshraghian, Alireza Firooz, Hassan Seirafi, Amir Hooshang Ehsani, Maryam Chamari, and Abbas Mirshafiey. "Randomized Controlled Trial Using Vitamins E and D Supplementation in Atopic Dermatitis." *Journal of Dermatological Treatment* 22, no. 3 (June 2011): 144–50. doi:10.3109/09546630903578566.

Jenike, Mark R. "Nutritional Ecology: Diet, Physical Activity and Body Size." In *Hunter-Gatherers: An Interdisciplinary Perspective*, edited by Catherine Panter-Brick, Robert H. Layton, and Peter Rowley-Conwy, 171–204. Cambridge University Press, 2001.

Jha, Durgesh Nandan. "Man Dies After Drinking Lauki Juice." *Times of India*. July 10, 2010. http://articles.timesofindia.indiatimes.com/2010-07-10/delhi /28310996_1_juice-bitter-taste-gourd.

Johannesson, Elisabet, Magnus Simrén, Hans Strid, Antal Bajor, and Riadh Sa-dik. "Physical Activity Improves Symptoms in Irritable Bowel Syndrome: A Randomized Controlled Trial." *American Journal of Gastroenterology* 106, no. 5 (2011): 915–22.

Johnson, Richard J., Eric A. Gaucher, Yuri Y. Sautin, George N. Henderson, Alex J. Angerhofer, and Steven A. Benner. "The Planetary Biology of Ascorbate and Uric Acid and Their Relationship with the Epidemic of Obesity and Cardio-vascular Disease." *Medical Hypotheses* 71, no. 1 (2008): 22–31.

Jones, A. Maxwell P., Jerome A. Klun, Charles L. Cantrell, Diane Ragone, Kamlesh R. Chauhan, Paula N. Brown, and Susan J. Murch. "Isolation and Identification of Mosquito (*Aedes aegypti*) Biting Deterrent Fatty Acids from Male Inflo-rescences of Breadfruit (*Artocarpus altilis* [Parkinson] Fosberg)." *Journal of Agricultural and Food Chemistry* 60, no. 15 (2012): 3867–73.

Jones, Lisa A., Loraine T. Sinnott, Donald O. Mutti, Gladys L. Mitchell, Melvin L. Moeschberger, and Karla Zadnik. "Parental History of Myopia, Sports and Outdoor Activities, and Future Myopia." *Investigative Ophthalmology and Visual Science* 48, no. 8 (August 1, 2007): 3524–32. doi:10.1167/iovs.06-1118.

Jouan, Pierre-Nicolas, Yves Pouliot, Sylvie F. Gauthier, and Jean-Paul Laforest. "Hormones in Bovine Milk and Milk Products: A Survey." *International Dairy Journal* 16, no. 11 (November 2006): 1408–14. doi:10.1016/j.idairyj.2006.06.007.

Joyce, Colin. "Japanese Get a Taste for Western Food and Fall Victim to Obesity and Early Death." *Telegraph*, September 4, 2006. www.telegraph.co.uk/health /healthnews/3342882/Japanese-get-a-taste-for-Western-food-and-fall-victim -to-obesity-and-early-death.html.

Juul, Anders. "Serum Levels of Insulin-like Growth Factor I and Its Binding Pro-teins in Health and Disease." *Growth Hormone and IGF Research* 13, no. 4 (2003): 113–70.

Kafatos, Anthony, Hans Verhagen, Joanna Moschandreas, Ioanna Apostolaki, and Johannes J. M. Van Westerop. "Mediterranean Diet of Crete: Foods and Nu-trient Content." *Journal of the Academy of Nutrition and Dietetics* 100, no. 12 (December 2000): 1487–93.

Kapellakis, Iosif E., Konstantinos P. Tsagarakis, and John C. Crowther. "Olive Oil History, Production and By-Product Management." *Reviews in Environ-mental Science and Biotechnology* 7, no. 1 (2008): 1–26. doi:10.1007/s11157-007-9120-9.

Karmakar, R. N. *Forensic Medicine and Toxicology: Oral, Practical and MCQ.* 3rd ed. Academic Publishers, 2007.

Katz, S. H., M. L. Hediger, and L. A. Valleroy. "Traditional Maize Processing Techniques in the New World." *Science* 184, no. 4138 (May 17, 1974): 765–73. doi:10.2307/1738647.

Kenward, R. E., and J. L. Holm. "On the Replacement of the Red Squirrel in Britain: A Phytotoxic Explanation." *Proceedings of the Royal Society of London, Series B: Biological Sciences* 251, no. 1332 (March 22, 1993): 187–94. doi:10.1098 /rspb.1993.0028.

Kerr, George H. *Okinawa: The History of an Island People.* C. E. Tuttle, 1958.

Khan, Nguyen Cong, Ha Huy Tue, Bach Mai Le, Gia Vinh Le, and Ha Huy Khoi. "Secular Trends in Growth and Nutritional Status of Vietnamese Adults in Ru-ral Red River Delta after 30 Years (1976–2006)." *Asia Pacific Journal of Clinical Nutrition* 19, no. 3 (2010): 412.

Kinde, Hailu, Eileen Foate, Emily Beeler, Fransisco Uzal, Janet Moore, and Robert Poppenga. "Strong Circumstantial Evidence for Ethanol Toxicosis in Cedar Waxwings (*Bombycilla cedrorum*)." *Journal of Ornithology* 153, no. 3 (July 1, 2012): 995–98. doi:10.1007/s10336-012-0858-7.

King, Richard J. *Lobster.* Reaktion Books, 2012.

Kinney, Dennis K., Pamela Teixeira, Diane Hsu, Siena C. Napoleon, David J. Crowley, Andrea Miller, William Hyman, and Emerald Huang. "Relation of Schizophrenia Prevalence to Latitude, Climate, Fish Consumption, Infant Mortality, and Skin Color: A Role for Prenatal Vitamin D Deficiency and Infections?" *Schizophrenia Bulletin* 35, no. 3 (May 1, 2009): 582–95. doi:10.1093/schbul/sbp023.

Klugman, Keith P., Shabir A. Madhi, Robin E. Huebner, Robert Kohberger, Nontombi Mbelle, and Nathaniel Pierce. "A Trial of a 9-Valent Pneumococcal Conjugate Vaccine in Children with and Those Without HIV Infection." *New England Journal of Medicine* 349, no. 14 (October 2, 2003): 1341–48. doi:10.1056/NEJMoa035060.

Knapp, Gunnar, Cathy A. Roheim, and James Lavalette Anderson. *The Great Salmon Run: Competition Between Wild and Farmed Salmon.* TRAFFIC North America and World Wildlife Fund, 2007.

Kodama, Arthur M., and Yoshitsugi Hokama. "Variations in Symptomatology of Ciguatera Poisoning." *Toxicon* 27, no. 5 (1989): 593–95. doi:10.1016/0041-0101(89)90121-9.

Koh, Woon-Puay, Anna H. Wu, Renwei Wang, Li-Wei Ang, Derrick Heng, Jian-Min Yuan, and Mimi C. Yu. "Gender-Specific Associations Between Soy and Risk of Hip Fracture in the Singapore Chinese Health Study." *American Journal of Epidemiology* 170, no. 7 (October 1, 2009): 901–9. doi:10.1093/aje/kwp220.

Kopicki, Allison. "Strong Support for Labeling Modified Foods." *New York Times,* July 27, 2013. www.nytimes.com/2013/07/28/science/strong-support-for-labeling-modified-foods.html.

Kowalchik, Claire, and William H. Hylton. *Rodale's Illustrated Encyclopedia of Herbs.* Rodale, 1998.

Krakauer, Jon. "How Chris McCandless Died." *New Yorker,* September 12, 2013. www.newyorker.com/books/page-turner/how-chris-mccandless-died.

Kratzer, James T., Miguel A. Lanaspa, Michael N. Murphy, Christina Cicerchi, Christina L. Graves, Peter A. Tipton, Eric A. Ortlund, Richard J. Johnson, and Eric A. Gaucher. "Evolutionary History and Metabolic Insights of Ancient Mammalian Uricases." *Proceedings of the National Academy of Sciences* 111, no. 10 (March 11, 2014): 3763–68. doi:10.1073/pnas.1320393111.

Kretchmer, N., R. Hurwitz, O. Ransome-Kuti, C. Dungy, and W. Alakija. "Intestinal Absorption of Lactose in Nigerian Ethnic Groups." *Lancet* 298, no. 7721 (1971): 392–95.

Kulick, Don, and Anne Meneley. *Fat: The Anthropology of an Obsession*. Jeremy P. Tarcher/Penguin, 2005.

Kull, I., A. Bergström, G. Lilja, G. Pershagen, and M. Wickman. "Fish Consumption During the First Year of Life and Development of Allergic Diseases During Childhood." *Allergy* 61, no. 8 (2006): 1009–15. doi:10.1111/j.1398-9995.2006.01115.x.

Kurian, A., and M. Asha Sankar. *Medicinal Plants*, vol. 2. Horticulture Science Series. New India Publishing, 2007.

Kurihara, Kenzo. "Glutamate: From Discovery as a Food Flavor to Role as a Basic Taste (Umami)." *American Journal of Clinical Nutrition* 90, no. 3 (September 1, 2009): 719S–722S. doi:10.3945/ajcn.2009.27462D.

Kynaston, H. G., D. I. Lewis-Jones, R. V. Lynch, and A. D. Desmond. "Changes in Seminal Quality Following Oral Zinc Therapy." *Andrologia* 20, no. 1 (1988): 21–22.

Larsen, Clark Spencer. "The Agricultural Revolution as Environmental Catastrophe: Implications for Health and Lifestyle in the Holocene." *Quaternary International* 150, no. 1 (2006): 12–20.

Laskaris, George. *Color Atlas of Oral Diseases*. Thieme, 2003.

Lawler, Dennis F., Brian T. Larson, Joan M. Ballam, Gail K. Smith, Darryl N. Biery, Richard H. Evans, Elizabeth H. Greeley, Mariangela Segre, Howard D. Stowe, and Richard D. Kealy. "Diet Restriction and Ageing in the Dog: Major Observations over Two Decades." *British Journal of Nutrition* 99, no. 4 (December 6, 2007). doi:10.1017/S0007114507871686.

Lawley, Richard, Laurie Curtis, and Judy Davis. *The Food Safety Hazard Guidebook*. Royal Society of Chemistry, 2012.

Lazenby, Richard A., and Peter McCormack. "Salmon and Malnutrition on the Northwest Coast." *Current Anthropology* 26, no. 3 (June 1, 1985): 379–84. doi:10.2307/2742736.

"Leader in Healthcare & Preventive Medicine: Dean Ornish, MD." Accessed March 2, 2015. http://deanornish.com/about/.

Le Bourg, Eric. "About the Article 'Exploring the Impact of Climate on Human Longevity' (Exp. Geront. 47, 660–671, 2012)." *Experimental Gerontology*, October 25, 2012. doi:10.1016/j.exger.2012.10.005.

Lehane, Leigh, and Richard J. Lewis. "Ciguatera: Recent Advances but the Risk Remains." *International Journal of Food Microbiology* 61, nos. 2–3 (November 1, 2000): 91–125. doi:10.1016/S0168-1605(00)00382-2.

Leibowitz, U., A. Antonovsky, J. M. Medalie, H. A. Smith, L. Halpern, and M. Alter. "Epidemiological Study of Multiple Sclerosis in Israel, II: Multiple

Sclerosis and Level of Sanitation." *Journal of Neurology, Neurosurgery, and Psychiatry* 29, no. 1 (1966): 60.

Leigh, G. J. *The World's Greatest Fix: A History of Nitrogen and Agriculture.* Oxford University Press, 2004.

Lemon, P. W. "Effects of Exercise on Dietary Protein Requirements." *International Journal of Sport Nutrition* 8, no. 4 (1998): 426–47.

Leonardi, Michela, Pascale Gerbault, Mark G. Thomas, and Joachim Burger. "The Evolution of Lactase Persistence in Europe. A Synthesis of Archaeological and Genetic Evidence." *International Dairy Journal* 22, no. 2 (February 2012): 88–97. doi:10.1016/j.idairyj.2011.10.010.

Lepre, Christopher J., Hélène Roche, Dennis V. Kent, Sonia Harmand, Rhonda L. Quinn, Jean-Philippe Brugal, Pierre-Jean Texier, Arnaud Lenoble, and Craig S. Feibel. "An Earlier Origin for the Acheulian." *Nature* 477, no. 7362 (2011): 82–85.

Leshem, Micah, Amany Saadi, Nesreen Alem, and Khadeja Hendi. "Enhanced Salt Appetite, Diet and Drinking in Traditional Bedouin Women in the Negev." *Appetite* 50, no. 1 ( January 2008): 71–82. doi:10.1016/j.appet.2007.05.010.

Lesser, Lenard I., Cara B. Ebbeling, Merrill Goozner, David Wypij, and David S. Ludwig. "Relationship Between Funding Source and Conclusion Among Nutrition-Related Scientific Articles." *PLoS Medicine* 4, no. 1 ( January 9, 2007): e5. doi:10.1371/journal.pmed.0040005.

Levey, D. J., J. J. Tewksbury, I. Izhaki, E. Tsahar, and D. C. Haak. "Evolutionary Ecology of Secondary Compounds in Ripe Fruit: Case Studies with Capsaicin and Emodin." *Seed Dispersal: Theory and Its Application in a Changing World*, edited by A. J. Dennis, E. W. Schupp, R. J. Green, and D. A. Westcott, 37–58. CABI, 2007.

Levey, Douglas J. "The Evolutionary Ecology of Ethanol Production and Alcoholism." *Integrative and Comparative Biology* 44, no. 4 (2004): 284–89.

Levey, Douglas J., and Carlos Martínez del Rio. "It Takes Guts (and More) to Eat Fruit: Lessons from Avian Nutritional Ecology." *Auk* 118, no. 4 (2001): 819–31.

Lewis, R. A. *Lewis' Dictionary of Toxicology.* Informa HealthCare, 1998.

Li, Y. H., and S. P. Chen. "Evolutionary History of Ebola Virus." *Epidemiology and Infection* 142, no. 6 (2014): 1138–45.

Liberski, Pawel P., Beata Sikorska, Shirley Lindenbaum, Lev G. Goldfarb, Catriona McLean, Johannes A. Hainfellner, and Paul Brown. "Kuru: Genes, Cannibals and Neuropathology." *Journal of Neuropathology and Experimental Neurology* 71, no. 2 (2012): 92–103.

Libert, Bo, and Vincent R. Franceschi. "Oxalate in Crop Plants." *Journal of Agri-*

*cultural and Food Chemistry* 35, no. 6 (November 1, 1987): 926–38. doi:10.1021/jf00078a019.

Liebenberg, Louis. "Persistence Hunting by Modern Hunter-Gatherers." *Current Anthropology* 47, no. 6 (2006): 1017–26.

Linos, Eleni, Walter C. Willett, Eunyoung Cho, and Lindsay Frazier. "Adolescent Diet in Relation to Breast Cancer Risk Among Premenopausal Women." *Cancer Epidemiology Biomarkers and Prevention* 19, no. 3 (March 1, 2010): 689–96. doi:10.1158/1055-9965.EPI-09-0802.

Lionis, C., M. Bathianaki, N. Antonakis, S. Papavasiliou, and A. Philalithis. "A High Prevalence of Diabetes Mellitus in a Municipality of Rural Crete, Greece." *Diabetic Medicine* 18, no. 9 (2001): 768–69.

Liu, Hau, Dena M. Bravata, Ingram Olkin, Smita Nayak, Brian Roberts, Alan M. Garber, and Andrew R. Hoffman. "Systematic Review: The Safety and Efficacy of Growth Hormone in the Healthy Elderly." *Annals of Internal Medicine* 146, no. 2 (January 16, 2007): 104–15.

Livi-Bacci, Massimo. "Fertility, Nutrition, and Pellagra: Italy During the Vital Revolution." *Journal of Interdisciplinary History*, 1986, 431–54.

Lott, J. N. A., I. Ockenden, V. Raboy, and G. D. Batten. "Phytic Acid and Phosphorus in Crop Seeds and Fruits: A Global Estimate." *Seed Science Research* 10, no. 1 (2000): 11.

Luca, F., G. H. Perry, and A. Di Rienzo. "Evolutionary Adaptations to Dietary Changes." *Annual Review of Nutrition* 30 (2010): 291–314.

Ludy, Mary-Jon, George E. Moore, and Richard D. Mattes. "The Effects of Capsaicin and Capsiate on Energy Balance: Critical Review and Meta-Analyses of Studies in Humans." *Chemical Senses* 37, no. 2 (2012): 103–21. doi:10.1093/chemse/bjr100.

Lund, E. M., P. J. Armstrong, Claudia A. Kirk, and J. S. Klausner. "Prevalence and Risk Factors for Obesity in Adult Cats from Private US Veterinary Practices." *International Journal of Applied Research in Veterinary Medicine* 3, no. 2 (2005): 88–96.

MacLean, C. H., S. J. Newberry, W. A. Mojica, P. Khanna, A. M. Issa, M. J. Suttorp, Y. W. Lim, S. B. Traina, L. Hilton, and R. Garland. "Effects of Omega-3 Fatty Acids on Cancer Risk." *JAMA: The Journal of the American Medical Association* 295, no. 4 (2006): 403–15.

"Madsen, Ole, Mark Scally, Christophe J. Douady, Diana J. Kao, Ronald W. DeBry, Ronald Adkins, Heather M. Amrine, Michael J. Stanhope, Wilfried W. de Jong, and Mark S. Springer. "Parallel Adaptive Radiations in Two Major Clades of Placental Mammals." *Nature* 409, no. 6820 (February 1, 2001): 610–14. doi: 10.1038/35054544.

Mahoney, Shane Patrick. "Recreational Hunting and Sustainable Wildlife Use in North America." In *Recreational Hunting, Conservation and Rural Livelihoods*, edited by Barney Dickson, Jon Hutton, and William M. Adams, 266–81. Wiley-Blackwell, 2009. http://onlinelibrary.wiley.com/doi/10.1002/9781444303179 .ch16/summary.

Maji, Debasish. "Vitamin D Toxicity." *Indian Journal of Endocrinology and Metabolism* 16, no. 2 (2012): 295.

Malacarne, Massimo. "Protein and Fat Composition of Mare's Milk: Some Nutritional Remarks with Reference to Human and Cow's Milk." *International Dairy Journal* 12, no. 11 (2002): 869–77.

Malainey, M. E., R. Przybylski, and B. L. Sherriff. "One Person's Food: How and Why Fish Avoidance May Affect the Settlement and Subsistence Patterns of Hunter-Gatherers." *American Antiquity* 66, no. 1 (January 1, 2001): 141–61. doi:10.2307/2694322.

Marcus, Gary. *Kluge.* Houghton Mifflin Harcourt, 2009.

Mariani-Costantini, R., and A. Mariani-Costantini. "An Outline of the History of Pellagra in Italy." *Journal of Anthropological Sciences* 85 (2007): 163–71.

Marlowe, Frank W. "Hunter-Gatherers and Human Evolution." *Evolutionary Anthropology: Issues, News, and Reviews* 14, no. 2 (2005): 54–67.

Marmot, M. G. "Alcohol and Coronary Heart Disease." *International Journal of Epidemiology* 13, no. 2 (June 1, 1984): 160–67. doi:10.1093/ije/13.2.160.

Marmot, Michael G. "Commentary: Reflections on Alcohol and Coronary Heart Disease." *International Journal of Epidemiology* 30, no. 4 (August 1, 2001): 729–34. doi:10.1093/ije/30.4.729.

Martial. *Epigrams.* Loeb Classical Library edition. Harvard University Press, 1993.

Mattison, Julie A., George S. Roth, T. Mark Beasley, Edward M. Tilmont, April M. Handy, Richard L. Herbert, Dan L. Longo, et al. "Impact of Caloric Restriction on Health and Survival in Rhesus Monkeys from the NIA Study." *Nature* 489, no. 7415 (September 13, 2012): 318–21. doi:10.1038/nature11432.

Mattson, Mark P., and Ruiqian Wan. "Beneficial Effects of Intermittent Fasting and Caloric Restriction on the Cardiovascular and Cerebrovascular Systems." *Journal of Nutritional Biochemistry* 16, no. 3 (2005): 129–37.

May, John F. *World Population Policies: Their Origin, Evolution, and Impact.* Springer Science & Business Media, 2012.

McGovern, Patrick E., Juzhong Zhang, Jigen Tang, Zhiqing Zhang, Gretchen R. Hall, Robert A. Moreau, Alberto Nuñez, et al. "Fermented Beverages of Pre- and Proto-Historic China." *Proceedings of the National Academy of Sciences* 101, no. 51 (2004): 17593–98.

McGreevy, P. D., P. C. Thomson, C. Pride, A. Fawcett, T. Grassi, and B. Jones.

"Prevalence of Obesity in Dogs Examined by Australian Veterinary Practices and the Risk Factors Involved." *Veterinary Record* 156, no. 22 (May 28, 2005): 695–702.

McMillan, Mary, and J. C. Thompson. "An Outbreak of Suspected Solanine Poisoning in Schoolboys: Examination of Criteria of Solanine Poisoning." *QJM* 48, no. 2 (April 1, 1979): 227–43.

Mead, Simon, Michael P. H. Stumpf, Jerome Whitfield, Jonathan A. Beck, Mark Poulter, Tracy Campbell, James B. Uphill, et al. "Balancing Selection at the Prion Protein Gene Consistent with Prehistoric Kurulike Epidemics." *Science* 300, no. 5619 (2003): 640–43.

Melnik, Bodo C. "Milk—The Promoter of Chronic Western Diseases." *Medical Hypotheses* 72, no. 6 (2009): 631–39.

Meng, Weihua, Jacqueline Butterworth, Patrick Calvas, and Francois Malecaze. "Myopia and Iris Colour: A Possible Connection?" *Medical Hypotheses* 78, no. 6 (June 2012): 778–80. doi:10.1016/j.mehy.2012.03.005.

Mensink, Ronald P., Peter L. Zock, Arnold DM Kester, and Martijn B. Katan. "Effects of Dietary Fatty Acids and Carbohydrates on the Ratio of Serum Total to HDL Cholesterol and on Serum Lipids and Apolipoproteins: A Meta-Analysis of 60 Controlled Trials." *American Journal of Clinical Nutrition* 77, no. 5 (May 1, 2003): 1146–55.

Meuninck, Jim. *Medicinal Plants of North America: A Field Guide*. Globe Pequot, 2008.

Meyer, Barbara J., Neil J. Mann, Janine L. Lewis, Greg C. Milligan, Andrew J. Sinclair, and Peter R. C. Howe. "Dietary Intakes and Food Sources of Omega-6 and Omega-3 Polyunsaturated Fatty Acids." *Lipids* 38, no. 4 (April 1, 2003): 391–98. doi:10.1007/s11745-003-1074-0.

Meyer, Martin Willy, and Jacob P. Thyssen. "Filaggrin Gene Defects and Dry Skin Barrier Function." In *Treatment of Dry Skin Syndrome*, edited by Marie Lodén and Howard I. Maibach, 119–24. Springer-Verlag Berlin Heidelberg, 2012. http://link.springer.com.proxy.bib.uottawa.ca/chapter/10.1007/978-3-642-27606-4_9.

Miller, Gregory D., Judith K. Jarvis, and Lois D. McBean. *Handbook of Dairy Foods and Nutrition,* Third Edition. CRC Press, 2006.

Miller, Holly C., and Charlotte Bender. "The Breakfast Effect: Dogs (*Canis familiaris*) Search More Accurately When They Are Less Hungry." *Behavioural Processes* 91, no. 3 (November 2012): 313–17. doi:10.1016/j.beproc.2012.09.012.

Miller, Kirk, Berwood Yost, Sean Flaherty, Marianne M. Hillemeier, Gary A. Chase, Carol S. Weisman, and Anne-Marie Dyer. "Health Status, Health Conditions, and Health Behaviors Among Amish Women: Results from the Central Pennsylvania Women's Health Study (CePAWHS)." *Women's Health Issues* 17, no. 3 (May 2007): 162–71. doi:10.1016/j.whi.2007.02.011.

Mintz, Sidney W., and Daniela Schlettwein-Gsell. "Food Patterns in Agrarian Societies: The 'Core-Fringe-Legume Hypothesis'—A Dialogue." *Gastronomica* 1, no. 3 (Summer 2001): 40–52. doi:10.1525/gfc.2001.1.3.40.

Mithen, Steven. "'Whatever Turns You On': A Response to Anna Machin,'Why Handaxes Just Aren't That Sexy.'" *Antiquity* 82, no. 317 (2008): 766–69.

Moher, D., H. M. Schachter, J. Reisman, K. Tran, B. Dales, K. Kourad, D. Barnes, M. Sampson, Andra Morrison, Isabelle Gaboury, and Janine Blackman. *Health Effects of Omega-3 Fatty Acids on Asthma.* Prepared for U.S. Department of Health and Human Services, Agency for Healthcare Research and Quality, 2004. http://internet.ahrq.gov/downloads/pub/evidence/pdf/o3asthma/o3asthma.pdf.

Molasky, Michael S. *The American Occupation of Japan and Okinawa: Literature and Memory.* Psychology Press, 1999.

Monarca, Silvano, Frantisek Kozisek, Gunther Craun, Francesco Donato, and Maria Zerbini. "Drinking Water Hardness and Cardiovascular Disease." *European Journal of Cardiovascular Prevention and Rehabilitation* 16, no. 6 (2009): 735–36.

Mosby, Ian. "'That Won-Ton Soup Headache': The Chinese Restaurant Syndrome, MSG and the Making of American Food, 1968–1980." *Social History of Medicine* 22, no. 1 (2009): 133–51.

Motley, Timothy J., Nyree Zerega, and Hugh B. Cross, eds. *Darwin's Harvest: New Approaches to the Origin, Evolution and Conservation of Crops.* Columbia University Press, 2006.

Mulcahy, Grace, Sandra O'Neill, June Fanning, Elaine McCarthy, and Mary Sekiya. "Tissue Migration by Parasitic Helminths—An Immunoevasive Strategy?" *Trends in Parasitology* 21, no. 6 (June 2005): 273–77. doi:10.1016/j.pt.2005.04.003.

Mullins, Raymond J., Sunday Clark, and Carlos A. Camargo Jr. "Regional Variation in Infant Hypoallergenic Formula Prescriptions in Australia." *Pediatric Allergy and Immunology* 21, no. 2p2 (2010): e413–20. doi:10.1111/j.1399-3038.2009.00962.x.

———. "Regional Variation in Epinephrine Autoinjector Prescriptions in Australia: More Evidence for the Vitamin D–anaphylaxis Hypothesis." *Annals of Allergy, Asthma and Immunology* 103, no. 6 (December 2009): 488–95. doi:10.1016/S1081-1206(10)60265-7.

Munro, Natalie. "Epipaleolithic Subsistence Intensification in the Southern Levant: The Faunal Evidence." In *The Evolution of Hominin Diets*, edited by Jean-Jacques Hublin and Michael P. Richards, 141–55. Springer, 2009. http://link.springer.com/chapter/10.1007/978-1-4020-9699-0_10.

Murgatroyd, Sarah. *Dig 3ft NW: The Legendary Journey of Burke and Wills.* Text Publishing, 2010.

————. *The Dig Tree: A True Story of Bravery, Insanity, and the Race to Discover Australia's Wild Frontier.* Random House Digital, 2002.

Murray, Stuart. *Atlas of American Military History.* Infobase Publishing, 2004.

Nafstad, Per, Wenche Nystad, Per Magnus, and Jouni J. K. Jaakkola. "Asthma and Allergic Rhinitis at 4 Years of Age in Relation to Fish Consumption in Infancy." *Journal of Asthma* 40, no. 4 (2003): 343–48. http://informahealthcare.com/doi/abs/10.1081/JAS-120018633.

Nakagawa, S., M. Lagisz, K. L. Hector, and H. G. Spencer. "Comparative and Meta-Analytic Insights into Life Extension via Dietary Restriction." *Aging Cell* 11, no. 3 ( June 2012): 401–9. http://onlinelibrary.wiley.com/doi/10.1111/j.1474-9726.2012.00798.x/full.

Nakamura, Yasuyuki, Hirotsugu Ueshima, Tomonori Okamura, Takashi Kadowaki, Takehito Hayakawa, Yoshikuni Kita, Robert D. Abbott, and Akira Okayama. "A Japanese Diet and 19-Year Mortality: National Integrated Project for Prospective Observation of Non-Communicable Diseases and Its Trends in the Aged, 1980." *British Journal of Nutrition* 101, no. 11 (2009): 1696–705. doi:10.1017/S0007114508111503.

Nellis, David W. *Poisonous Plants and Animals of Florida and the Caribbean.* Pineapple Press, 1997.

Newcomb, T. P., and R. T. Spurr. *A Technical History of the Motor Car.* A. Hilger, 1989.

Nielsen. *State of the Media TV Usage Trends: Q2 2010.* www.nielsen.com/content/dam/corporate/us/en/newswire/uploads/2010/11/Nielsen-Q2-2010-State-of-the-Media-Fact-Sheet.pdf.

Nieminen, M. S., M. P. Rämö, M. Viitasalo, P. Heikkilä, J. Karjalainen, M. Mäntysaari, and J. Heikkila. "Serious Cardiovascular Side Effects of Large Doses of Anabolic Steroids in Weight Lifters." *European Heart Journal* 17, no. 10 (October 1, 1996): 1576–83.

Nimptsch, Katharina, Sabine Rohrmann, Rudolf Kaaks, and Jakob Linseisen. "Dietary Vitamin K Intake in Relation to Cancer Incidence and Mortality: Results from the Heidelberg Cohort of the European Prospective Investigation into Cancer and Nutrition (EPIC-Heidelberg)." *American Journal of Clinical Nutrition* 91, no. 5 (May 1, 2010): 1348–58. doi:10.3945/ajcn.2009.28691.

Njoku, P. C., A. A. Ayuk, and C. V. Okoye. "Temperature Effects on Vitamin C Content in Citrus Fruits." *Pakistan Journal of Nutrition* 10, no. 12 (December 1, 2011): 1168–69. doi:10.3923/pjn.2011.1168.1169.

Norton, C. J., Y. Kondo, A. Ono, Y. Zhang, and M. C. Diab. "The Nature of Megafaunal Extinctions During the MIS 3–2 Transition in Japan." *Quaternary International* 211, no. 1 (2010): 113–22.

Norton, Heather L., Rick A. Kittles, Esteban Parra, Paul McKeigue, Xianyun

Mao, Keith Cheng, Victor A. Canfield, Daniel G. Bradley, Brian McEvoy, and Mark D. Shriver. "Genetic Evidence for the Convergent Evolution of Light Skin in Europeans and East Asians." *Molecular Biology and Evolution* 24, no. 3 (March 1, 2007): 710–22. doi:10.1093/molbev/msl203.

Oonincx, Dennis G. A. B., Joost van Itterbeeck, Marcel J. W. Heetkamp, Henry van den Brand, Joop J. A. van Loon, and Arnold van Huis. "An Exploration on Greenhouse Gas and Ammonia Production by Insect Species Suitable for Animal or Human Consumption." *PLoS ONE* 5, no. 12 (December 29, 2010): e14445. doi:10.1371/journal.pone.0014445.

Osawa, Rinko, Satoshi Konno, Masashi Akiyama, Ikue Nemoto-Hasebe, Toshifumi Nomura, Yukiko Nomura, Riichiro Abe, et al. "Japanese-Specific Filaggrin Gene Mutations in Japanese Patients Suffering from Atopic Eczema and Asthma." *Journal of Investigative Dermatology* 130, no. 12 (2010): 2834–36. doi:10.1038/jid.2010.218.

Osborne, Nicholas J., Jennifer J. Koplin, Pamela E. Martin, Lyle C. Gurrin, Adrian J. Lowe, Melanie C. Matheson, Anne-Louise Ponsonby, Melissa Wake, Mimi L. K. Tang, Shyamali C. Dharmage, and Katrina J. Allen. "Prevalence of Challenge-Proven IgE-Mediated Food Allergy Using Population-Based Sampling and Predetermined Challenge Criteria in Infants." *Journal of Allergy and Clinical Immunology* 127, no. 3 (March 2011): 668–76.e2. doi:10.1016/j.jaci.2011.01.039.

Osborne, Nicholas J., Obioha C. Ukoumunne, Melissa Wake, and Katrina J. Allen. "Prevalence of Eczema and Food Allergy Is Associated with Latitude in Australia." *Journal of Allergy and Clinical Immunology* 129, no. 3 (March 2012): 865–67. doi:10.1016/j.jaci.2012.01.037.

Ostbye, T., T. J. Welby, I. A. Prior, C. E. Salmond, and Y. M. Stokes. "Type 2 (Non-Insulin-Dependent) Diabetes Mellitus, Migration and Westernisation: The Tokelau Island Migrant Study." *Diabetologia* 32, no. 8 (August 1989): 585–90.

Ottaviani, Monica, Emanuela Camera, and Mauro Picardo. "Lipid Mediators in Acne." *Mediators of Inflammation* 2010 (2010). http://www.hindawi.com/journals/mi/2010/858176/abs/.

Ou, Chun-Quan, Yun-Feng Song, Jun Yang, Patsy Yuen-Kwan Chau, Lin Yang, Ping-Yan Chen, and Chit-Ming Wong. "Excess Winter Mortality and Cold Temperatures in a Subtropical City, Guangzhou, China." *PLoS ONE* 8, no. 10 (2013): e77150.

Packard, Vernal S., Jr. *Processed Foods and the Consumer: Additives, Labeling, Standards, and Nutrition.* University of Minnesota Press, 1976.

Pálsson, Gísli. *Coastal Economies, Cultural Accounts: Human Ecology and Icelandic Discourse.* Manchester University Press, 1994.

Panda, H. *Herbs Cultivation and Medicinal Uses.* 2nd ed. National Institute of Industrial Research, 2000.

Paoletti, Maurizio G., Lorenzo Norberto, Roberta Damini, and Salvatore Musumeci. "Human Gastric Juice Contains Chitinase That Can Degrade Chitin." *Annals of Nutrition and Metabolism* 51, no. 3 (2007): 244–51. doi: 10.1159/000104144.

"Papua New Guinea." *Ethnologue.* Accessed August 20, 2014. www.ethnologue.com /country/PG/default/%2A%2A%2AEDITION%2A%2A%2A.

Parra, Esteban J. "Human Pigmentation Variation: Evolution, Genetic Basis, and Implications for Public Health." *American Journal of Physical Anthropology* 134, no. S45 (2007): 85–105. doi:10.1002/ajpa.20727.

Pearce, Neil, Nadia Aït-Khaled, Richard Beasley, Javier Mallol, Ulrich Keil, Ed Mitchell, Colin Robertson, and the ISAAC Phase Three Study Group. "Worldwide Trends in the Prevalence of Asthma Symptoms: Phase III of the International Study of Asthma and Allergies in Childhood (ISAAC)." *Thorax* 62, no. 9 (September 1, 2007): 758–66. doi:10.1136/thx.2006.070169.

Pella, D., G. Dubnov, R. B. Singh, R. Sharma, E. M. Berry, and O. Manor. "Effects of an Indo-Mediterranean Diet on the Omega-6/Omega-3 Ratio in Patients at High Risk of Coronary Artery Disease: The Indian Paradox." *World Review of Nutrition and Dietetics* 92 (2003): 74–80. http://content.karger.com /ProdukteDB/produkte.asp?Doi=73793.

Peng, Yi, Hong Shi, Xue-bin Qi, Chun-jie Xiao, Hua Zhong, Run-lin Z. Ma, and Bing Su. "The ADH1B Arg47His Polymorphism in East Asian Populations and Expansion of Rice Domestication in History." *BMC Evolutionary Biology* 10, no. 1 (2010): 15.

Perelman, Polina, Warren E. Johnson, Christian Roos, Hector N. Seuánez, Julie E. Horvath, Miguel A. M. Moreira, Bailey Kessing, Joan Pontius, Melody Roelke, Yves Rumpler, Maria Paula C. Schneider, Artur Silva, Stephen J. O'Brien, and Jill Pecon-Slattery. "A Molecular Phylogeny of Living Primates." *PLoS Genetics* 7, no. 3 (March 17, 2011): e1001342. doi:10.1371/journal.pgen.1001342.

Pérez-Jiménez, Francisco, Juan Ruano, Pablo Perez-Martinez, Fernando Lopez-Segura, and Jose Lopez-Miranda. "The Influence of Olive Oil on Human Health: Not a Question of Fat Alone." *Molecular Nutrition and Food Research* 51, no. 10 (2007): 1199–1208. doi:10.1002/mnfr.200600273.

Peroni, D. G., G. L. Piacentini, E. Cametti, I. Chinellato, and A. L. Boner. "Correlation Between Serum 25-Hydroxyvitamin D Levels and Severity of Atopic Dermatitis in Children." *British Journal of Dermatology* 164, no. 5 (2011): 1078–82. doi:10.1111/j.1365-2133.2010.10147.x.

Perry, George H., Nathaniel J. Dominy, Katrina G. Claw, Arthur S. Lee, Heike Fiegler, Richard Redon, John Werner, Fernando A. Villanea, Joanna L. Mountain, Rajeev Misra, Nigel P. Carter, Charles Lee, and Anne C. Stone. "Diet and the Evolution of Human Amylase Gene Copy Number Variation." *Nature Genetics* 39, no. 10 (September 9, 2007): 1256–60. doi:10.1038/ng2123.

Peters, H. P. F., W. R. De Vries, G. P. Vanberge-Henegouwen, and L. M. A. Akkermans. "Potential Benefits and Hazards of Physical Activity and Exercise on the Gastrointestinal Tract." *Gut* 48, no. 3 (2001): 435–39.

Piškur, Jure, Elżbieta Rozpędowska, Silvia Polakova, Annamaria Merico, and Concetta Compagno. "How Did *Saccharomyces* Evolve to Become a Good Brewer?" *Trends in Genetics* 22, no. 4 (2006): 183–86.

Plutarch. *Isis and Osiris.* In *Moralia,* Loeb Classical Library edition, vol. 5 (1936). http://penelope.uchicago.edu/Thayer/E/Roman/Texts/Plutarch/Moralia/Isis_and_Osiris***/E.html.

Pohanish, Richard P. *Sittig's Handbook of Toxic and Hazardous Chemicals and Carcinogens.* William Andrew, 2011.

Pollan, Michael. *Food Rules: An Eater's Manual.* Penguin, 2009.

Porth, Carol M. *Essentials of Pathophysiology: Concepts of Altered Health States.* Lippincott Williams & Wilkins, 2010.

Poulain, Michel, Giovanni Mario Pes, Claude Grasland, Ciriaco Carru, Luigi Ferrucci, Giovannella Baggio, Claudio Franceschi, and Luca Deiana. "Identification of a Geographic Area Characterized by Extreme Longevity in the Sardinia Island: The AKEA Study." *Experimental Gerontology* 39, no. 9 (September 2004): 1423–29. doi:10.1016/j.exger.2004.06.016.

Powles, John W., and D. Ruth. "Diet–Mortality Associations." *Medical Practice of Preventive Nutrition.* London: Smith–Gordon, 1994, 75–90.

Premalatha, M., Tasneem Abbasi, Tabassum Abbasi, and S. A. Abbasi. "Energy-Efficient Food Production to Reduce Global Warming and Ecodegradation: The Use of Edible Insects." *Renewable and Sustainable Energy Reviews* 15, no. 9 (2011): 4357–60.

Prentice, A. "Diet, Nutrition and the Prevention of Osteoporosis." *Public Health Nutrition* 7, no. 1a (2004): 227–43.

Prescott, Susan, and Katrina J. Allen. "Food Allergy: Riding the Second Wave of the Allergy Epidemic." *Pediatric Allergy and Immunology* 22, no. 2 (2011): 155–60. doi:10.1111/j.1399-3038.2011.01145.x.

"Preservation of Health in the Japanese Navy and Army." *British Medical Journal* 1, no. 2368 (May 19, 1906): 1175–76. www.ncbi.nlm.nih.gov/pmc/articles/PMC2381360/.

Price, T. Douglas, and Ofer Bar-Yosef. "The Origins of Agriculture: New Data,

New Ideas: An Introduction to Supplement 4." *Current Anthropology* 52, no. S4 (October 1, 2011): S163–74. doi:10.1086/659964.

Price-Pottenger Nutrition Foundation. "Traditional Diets." Accessed August 8, 2014. http://ppnf.org/about/about-price-and-pottenger/dr-pottenger/traditional-diets/.

Qian, Hai. *Nematode Nicotinic Acetylcholine Receptors: A Single-Channel Study in Ascaris Suum and Caenorhabditis Elegans.* ProQuest, 2007.

Ragone, Diane. *Breadfruit: Artocarpus altilis (Parkinson) Fosberg—Promoting the Conservation and Use of Underutilized and Neglected Crops 10.* International Plant Genetic Resource Institute, 1997.

Rajakumar, K. "Pellagra in the United States: A Historical Perspective." *Southern Medical Journal* 93, no. 3 (2000): 272.

Raubenheimer, David, and Jessica M. Rothman. "Nutritional Ecology of Entomophagy in Humans and Other Primates." *Annual Review of Entomology* 58 (2013): 141–60.

Ravdin, Peter M., Kathleen A. Cronin, Nadia Howlader, Christine D. Berg, Rowan T. Chlebowski, Eric J. Feuer, Brenda K. Edwards, and Donald A. Berry. "The Decrease in Breast-Cancer Incidence in 2003 in the United States." *New England Journal of Medicine* 356, no. 16 (2007): 1670–74. doi:10.1056/NEJMsr070105.

Rea, Ramona L., Lilian U. Thompson, and David J. A. Jenkins. "Lectins in Foods and Their Relation to Starch Digestibility." *Nutrition Research* 5, no. 9 (September 1985): 919–29. doi:10.1016/S0271-5317(85)80105-6.

Read, A. F., and A. Skorping. "The Evolution of Tissue Migration by Parasitic Nematode Larvae." *Parasitology* 111 (September 1995): 359–71.

Reddy, N. R., and M. D. Pierson. "Reduction in Antinutritional and Toxic Components in Plant Foods by Fermentation." *Food Research International* 27, no. 3 (1994): 281–90. doi:10.1016/0963-9969(94)90096-5.

Reilly, Genevieve. "Malloy Signs State GMO Labeling Law in Fairfield." *Connecticut Post*, December 11, 2013. www.ctpost.com/news/article/Malloy-signs-state-GMO-labeling-law-in-Fairfield-5056120.php.

Renehan, A. G. "Insulin-like Growth Factor (IGF)-I, IGF Binding Protein-3, and Cancer Risk: Systematic Review and Meta-Regression Analysis." *Lancet* 363, no. 9418 (April 4, 2004): 1346–53.

Richards, Audrey Isabel. *Land, Labour and Diet in Northern Rhodesia: An Economic Study of the Bemba Tribe.* LIT Verlag Münster, 1995.

Richerson, Peter J., and Robert Boyd. "Built for Speed, Not for Comfort: Darwinian Theory and Human Culture." *History and Philosophy of the Life Sciences* 23 (2001): 425–65.

Ridley, J. "An Account of an Endemic Disease of Ceylon Entitled Berri Berri." *Dublin Hospital Reports and Communications in Medicine and Surgery* 2 (1818): 227–53.

Riemann, Hans P., and Dean O. Cliver, eds. *Foodborne Infections and Intoxications.* Academic Press, 2006.

Rimm, E. B., P. Williams, K. Fosher, M. Criqui, and M. J. Stampfer. "Moderate Alcohol Intake and Lower Risk of Coronary Heart Disease: Meta-Analysis of Effects on Lipids and Haemostatic Factors." *BMJ Clinical Research* 319, no. 7224 (December 11, 1999): 1523–28. doi:10.1136/bmj.319.7224.1523.

Roan, Shari. "A Slow Change of Heart." *Los Angeles Times,* May 5, 1996. http://articles.latimes.com/1996-05-05/news/ls-913_1_dean-ornish/3.

Robbins, Charles T., Jennifer K. Fortin, Karyn D. Rode, Sean D. Farley, Lisa A. Shipley, and Laura A. Felicetti. "Optimizing Protein Intake as a Foraging Strategy to Maximize Mass Gain in an Omnivore." *Oikos* 116, no. 10 (2007): 1675–82.

Roberts, Margaret. *Margaret Roberts' A–Z of Herbs.* Struik, 1920.

Roberts, R., and Z. Jacobs. "The Lost Giants of Tasmania." *Australasian Science* 29, no. 9 (2008): 14–17.

Roberts, R. G., T. F. Flannery, L. K. Ayliffe, H. Yoshida, J. M. Olley, G. J. Prideaux, G. M. Laslett, A. Baynes, M. A. Smith, and R. Jones. "New Ages for the Last Australian Megafauna: Continent-Wide Extinction about 46,000 Years Ago." *Science* 292, no. 5523 (2001): 1888–92.

Robson, Pamela, ed. *Great Australian Speeches: Words That Shaped a Nation.* Pier 9, 2009.

Rode, K. D., and C. T. Robbins. "Why Bears Consume Mixed Diets During Fruit Abundance." *Canadian Journal of Zoology* 78, no. 9 (2000): 1640–45.

Roebroeks, Wil, and Paola Villa. "On the Earliest Evidence for Habitual Use of Fire in Europe." *Proceedings of the National Academy of Sciences* 108, no. 13 (March 29, 2011): 5209–14. doi:10.1073/pnas.1018116108.

Rogers, Imogen S., Kate Northstone, David B. Dunger, Ashley R. Cooper, Andy R. Ness, and Pauline M. Emmett. "Diet Throughout Childhood and Age at Menarche in a Contemporary Cohort of British Girls." *Public Health Nutrition* 13, no. 12 (2010): 2052–63.

Rose, K. A., I. G. Morgan, J. Ip, A. Kifley, S. Huynh, W. Smith, and P. Mitchell. "Outdoor Activity Reduces the Prevalence of Myopia in Children." *Ophthalmology* 115, no. 8 (2008): 1279.

Rowland, Glovioell W., Gary G. Schwartz, Esther M. John, and Sue Ann Ingles. "Calcium Intake and Prostate Cancer Among African Americans: Effect Modification by Vitamin D Receptor Calcium Absorption Genotype." *Journal of Bone and Mineral Research* 27, no. 1 (2012): 187–94. doi:10.1002/jbmr.505.

Rozin, P., and D. Schiller. "The Nature and Acquisition of a Preference for Chili Pepper by Humans." *Motivation and Emotion* 4, no. 1 (1980): 77–101. http://psycnet .apa.org/psycinfo/1981-21337-001.

Rudders, Susan A., Janice A. Espinola, and Carlos A. Camargo Jr. "North-South Differences in US Emergency Department Visits for Acute Allergic Reactions." *Annals of Allergy, Asthma and Immunology* 104, no. 5 (May 2010): 413–16. doi: 10.1016/j.anai.2010.01.022.

Rutter, Jill, and Stephen Percy. "The Pulse That Maims." *New Scientist*, August 23, 1984.

Sadasivam, S., and B. Thayumanayan. *Molecular Host Plant Resistance to Pests*. CRC Press, 2003.

Sadeghirad, Behnam, Shahrzad Motaghipisheh, Fariba Kolahdooz, Mohammad J. Zahedi, and Ali A. Haghdoost. "Islamic Fasting and Weight Loss: A Systematic Review and Meta-Analysis." *Public Health Nutrition* 27 (2012): 1–11.

Saha, S., D. C. Chant, J. L. Welham, and J. J. McGrath. "The Incidence and Prevalence of Schizophrenia Varies with Latitude." *Acta Psychiatrica Scandinavica* 114, no. 1 (2006): 36–39. doi:10.1111/j.1600-0447.2005.00742.x.

Saisithi, P. "Traditional Fermented Fish: Fish Sauce Production." In *Fisheries Processing*, edited by A. M. Martin, 111–31. Springer, 1994. http://link.springer .com/chapter/10.1007/978-1-4615-5303-8_5.

Sakakibara, Ryuji, Kuniko Tsunoyama, Hiroyasu Hosoi, Osamu Takahashi, Megumi Sugiyama, Masahiko Kishi, Emina Ogawa, Hitoshi Terada, Tomoyuki Uchiyama, and Tomonori Yamanishi. "Influence of Body Position on Defecation in Humans." *LUTS: Lower Urinary Tract Symptoms* 2, no. 1 (2010): 16–21. doi:10.1111/j.1757-5672.2009.00057.x.

Saladie, Palmira, Rosa Huguet, Antonio Rodriguez-Hidalgo, Isabel Caceres, Montserrat Esteban-Nadal, Juan Luis Arsuaga, José María Bermúdez de Castro, and Eudald Carbonell. "Intergroup Cannibalism in the European Early Pleistocene: The Range Expansion and Imbalance of Power Hypotheses." *Journal of Human Evolution* 63, no. 5 (2012): 682–95.

Salam, Muhammad T., Yu-Fen Li, Bryan Langholz, and Frank D. Gilliland. "Maternal Fish Consumption During Pregnancy and Risk of Early Childhood Asthma." *Journal of Asthma* 42, no. 6 (January 1, 2005): 513–18.

Salimei, Elisabetta and Francesco Fantuz. "Equid Milk for Human Consumption." *International Dairy Journal* 24, no. 2 (June 1, 2012): 130–42.

Samuels, Adrienne. "The Toxicity/Safety of Processed Free Glutamic Acid (MSG): A Study in Suppression of Information." *Accountability in Research* 6, no. 4 (1999): 259–310.

Sapone, Anna, Julio C. Bai, Carolina Ciacci, Jernej Dolinsek, Peter H. R. Green,

Marios Hadjivassiliou, Katri Kaukinen, Kamran Rostami, David S Sanders, Michael Schumann, Reiner Ullrich, Danilo Villalta, Umberto Volta, Carlo Catassi, and Alessio Fasano. "Spectrum of Gluten-Related Disorders: Consensus on New Nomenclature and Classification." *BMC Medicine* 10, no. 1 (February 7, 2012): 13. doi:10.1186/1741-7015-10-13.

Satya, Santosh, Lalit M. Bal, Poonam Singhal, and S.N. Naik. "Bamboo Shoot Processing: Food Quality and Safety Aspect (a Review)." *Trends in Food Science and Technology* 21, no. 4 (April 2010): 181–89. doi:10.1016/j.tifs.2009.11.002.

Saunders, Karin A., Tim Raine, Anne Cooke, and Catherine E. Lawrence. "Inhibition of Autoimmune Type 1 Diabetes by Gastrointestinal Helminth Infection." *Infection and Immunity* 75, no. 1 (January 1, 2007): 397–407. doi:10.1128/IAI.00664-06.

Sausenthaler, Stefanie, Sibylle Koletzko, Beate Schaaf, Irina Lehmann, Michael Borte, Olf Herbarth, Andrea von Berg, H.-Erich Wichmann, and Joachim Heinrich. "Maternal Diet During Pregnancy in Relation to Eczema and Allergic Sensitization in the Offspring at 2 Y of Age." *American Journal of Clinical Nutrition* 85, no. 2 (February 1, 2007): 530–37.

Savage, G. P., L. Vanhanen, S. M. Mason, and A. B. Ross. "Effect of Cooking on the Soluble and Insoluble Oxalate Content of Some New Zealand Foods." *Journal of Food Composition and Analysis* 13, no. 3 (June 2000): 201–6. doi:10.1006/jfca.2000.0879.

Saw, Seang-Mei. "A Synopsis of the Prevalence Rates and Environmental Risk Factors for Myopia." *Clinical and Experimental Optometry* 86, no. 5 (2003): 289–94. doi:10.1111/j.1444-0938.2003.tb03124.x.

Saynor, R., D. Verel, and T. Gillott. "The Long-Term Effect of Dietary Supplementation with Fish Lipid Concentrate on Serum Lipids, Bleeding Time, Platelets and Angina." *Atherosclerosis* 50, no. 1 (1984): 3–10.

Schaub, Bianca, Roger Lauener, and Erika von Mutius. "The Many Faces of the Hygiene Hypothesis." *Journal of Allergy and Clinical Immunology* 117, no. 5 (May 2006): 969–77. doi:10.1016/j.jaci.2006.03.003.

Schmelzer, Gabriëlla Harriët, and Ameenah Gurib-Fakim. *Medicinal Plants 1.* PROTA, 2008.

Scholey, Andrew B., Susan Harper, and David O. Kennedy. "Cognitive Demand and Blood Glucose." *Physiology and Behavior* 73, no. 4 (2001): 585–92.

Schooling, C. Mary, Shiu Lun Au Yeung, Guy Freeman, and Benjamin J. Cowling. "The Effect of Statins on Testosterone in Men and Women, a Systematic Review and Meta-Analysis of Randomized Controlled Trials." *BMC Medicine* 11 (2013): 57. doi:http://dx.doi.org/10.1186/1741-7015-11-57.

Seigler, David Stanley. *Plant Secondary Metabolism.* Springer, 1998.

Sellers, Elizabeth A. C., Atul Sharma, and Celia Rodd. "Adaptation of Inuit Children to a Low-Calcium Diet." *Canadian Medical Association Journal* 168, no. 9 (April 29, 2003): 1141–43.

Semaw, Sileshi, Michael J. Rogers, Jay Quade, Paul R. Renne, Robert F. Butler, Manuel Dominguez-Rodrigo, Dietrich Stout, William S. Hart, Travis Pickering, and Scott W. Simpson. "2.6-Million-Year-Old Stone Tools and Associated Bones from OGS-6 and OGS-7, Gona, Afar, Ethiopia." *Journal of Human Evolution* 45, no. 2 (August 2003): 169–77. doi:10.1016/S0047-2484(03)00093-9.

Semba, Richard D., Emily J. Nicklett, and Luigi Ferrucci. "Does Accumulation of Advanced Glycation End Products Contribute to the Aging Phenotype?" *Journals of Gerontology Series A: Biological Sciences and Medical Sciences* 65A, no. 9 (September 1, 2010): 963–75. doi:10.1093/gerona/glq074.

Semba, Richard D., Luigi Ferrucci, Kai Sun, Justine Beck, Mansi Dalal, Ravi Varadhan, Jeremy Walston, Jack M. Guralnik, and Linda P. Fried. "Advanced Glycation End Products and Their Circulating Receptors Predict Cardiovascular Disease Mortality in Older Community-Dwelling Women." *Aging Clinical and Experimental Research* 21, no. 2 (April 2009): 182–90.

Serra-Majem, Lluís, Lourdes Ribas, Ricard Tresserras, Joy Ngo, and Lluís Salleras. "How Could Changes in Diet Explain Changes in Coronary Heart Disease Mortality in Spain? The Spanish Paradox." *American Journal of Clinical Nutrition* 61, no. 6 (1995): 1351S— 1359S.

Serrano, José, Riitta Puupponen-Pimiä, Andreas Dauer, Anna-Marja Aura, and Fulgencio Saura-Calixto. "Tannins: Current Knowledge of Food Sources, Intake, Bioavailability and Biological Effects." *Molecular Nutrition and Food Research* 53, no. S2 (2009): S310–29. doi:10.1002/mnfr.200900039.

Shanley, D. P., and T. B. L. Kirkwood. "Calorie Restriction and Aging: A Life-History Analysis." *Evolution* 54, no. 3 (2000): 740–50.

Shek, Lynette Pei-Chi, and Bee Wah Lee. "Food Allergy in Asia." *Current Opinion in Allergy and Clinical Immunology* 6, no. 3 (June 2006): 197–201. doi:10.1097/01.all.0000225160.52650.17.

Sherman, P. W., and J. Billing. "Darwinian Gastronomy: Why We Use Spices." *BioScience* 49, no. 6 (June 1999): 453–63.

Sherriff, A., and J. Golding. "Hygiene Levels in a Contemporary Population Cohort Are Associated with Wheezing and Atopic Eczema in Preschool Infants." *Archives of Disease in Childhood* 87, no. 1 (July 1, 2002): 26–29. doi:10.1136/adc.87.1.26.

Sherwin, Justin C., Alex W. Hewitt, Minas T. Coroneo, Lisa S. Kearns, Lyn R. Griffiths, and David A. Mackey. "The Association Between Time Spent Outdoors and Myopia Using a Novel Biomarker of Outdoor Light Exposure."

*Investigative Ophthalmology and Visual Science* 53, no. 8 (July 1, 2012): 4363–70. doi:10.1167/iovs.11-8677.

Shi, Peng, Jianzhi Zhang, Hui Yang, and Ya-ping Zhang. "Adaptive Diversification of Bitter Taste Receptor Genes in Mammalian Evolution." *Molecular Biology and Evolution* 20, no. 5 (May 1, 2003): 805–14. doi:10.1093/molbev/msg083.

Shimada, Akiko, Brian E. Cairns, Nynne Vad, Kathrine Ulriksen, Anne Marie Lynge Pedersen, Peter Svensson, and Lene Baad-Hansen. "Headache and Mechanical Sensitization of Human Pericranial Muscles After Repeated Intake of Monosodium Glutamate (MSG)." *Journal of Headache and Pain* 14, no. 1 (December 1, 2013): 1–9. doi:10.1186/1129-2377-14-2.

Sidbury, R., A. F. Sullivan, R. I. Thadhani, and C. A. Camargo Jr. "Randomized Controlled Trial of Vitamin D Supplementation for Winter-Related Atopic Dermatitis in Boston: A Pilot Study." *British Journal of Dermatology* 159, no. 1 (2008): 245–47. doi:10.1111/j.1365-2133.2008.08601.x.

Sieber, W. Karl, Cynthia F. Robinson, Jan Birdsey, Guang X. Chen, Edward M. Hitchcock, Jennifer E. Lincoln, Akinori Nakata, and Marie H. Sweeney. "Obesity and Other Risk Factors: The National Survey of US Long-Haul Truck Driver Health and Injury." *American Journal of Industrial Medicine* 57, no. 6 (2014): 615–26.

Siegel, Ronald K. *Intoxication: The Universal Drive for Mind-Altering Substances.* Inner Traditions/Bear, 2005.

Siemens, Jan, Sharleen Zhou, Rebecca Piskorowski, Tetsuro Nikai, Ellen A. Lumpkin, Allan I. Basbaum, David King, and David Julius. "Spider Toxins Activate the Capsaicin Receptor to Produce Inflammatory Pain." *Nature* 444, no. 7116 (November 9, 2006): 208–12. doi:10.1038/nature05285.

Siener, Roswitha, Ruth Hönow, Susanne Voss, Ana Seidler, and Albrecht Hesse. "Oxalate Content of Cereals and Cereal Products." *Journal of Agricultural and Food Chemistry* 54, no. 8 (April 1, 2006): 3008–11. doi:10.1021/jf052776v.

Sikirov, Dov. "Comparison of Straining During Defecation in Three Positions: Results and Implications for Human Health." *Digestive Diseases and Sciences* 48, no. 7 (2003): 1201–5.

Siler, Julia Flynn. "'Food of the Future' Has One Hitch: It's All but Inedible." *Wall Street Journal*, November 1, 2011. http://online.wsj.com/article/SB1000142405 2970203752604576645242121126386.html.

Silverberg, Jonathan I., Edward Kleiman, Nanette B. Silverberg, Helen G. Durkin, Rauno Joks, and Tamar A. Smith-Norowitz. "Chickenpox in Childhood Is Associated with Decreased Atopic Disorders, IgE, Allergic Sensitization, and Leukocyte Subsets." *Pediatric Allergy and Immunology* 23, no. 1 (2012): 50–58.

Simons, F. Estelle R., Sandra Peterson, and Charlyn D. Black. "Epinephrine Dispensing Patterns for an Out-of-Hospital Population: A Novel Approach to Studying the Epidemiology of Anaphylaxis." *Journal of Allergy and Clinical Immunology* 110, no. 4 (October 2002): 647–51. doi:10.1067/mai.2002.127860.

Simoons, Frederick J. *Eat Not This Flesh: Food Avoidances from Prehistory to the Present.* University of Wisconsin Press, 1994.

———. "Fish as Forbidden Food: The Case of India." *Ecology of Food and Nutrition* 3, no. 3 (1974): 185–201. doi:10.1080/03670244.1974.9990381.

———. "Rejection of Fish as Human Food in Africa: A Problem in History and Ecology." *Ecology of Food and Nutrition* 3, no. 2 (1974): 89–105. doi:10.1080/036 70244.1974.9990367.

Simopoulos, A. P. "The Importance of the Omega-6/Omega-3 Fatty Acid Ratio in Cardiovascular Disease and Other Chronic Diseases." *Experimental Biology and Medicine* 233, no. 6 (June 1, 2008): 674–88. doi:10.3181/0711-MR-311.

Singer, Joyce Z., and Stanley L. Wallace. "The Allopurinol Hypersensitivity Syndrome: Unnecessary Morbidity and Mortality." *Arthritis and Rheumatism* 29, no. 1 (1986): 82–87.

Singletary, Keith. "Red Pepper: Overview of Potential Health Benefits." *Nutrition Today* 46, no. 1 (2011): 33–47. doi:10.1097/NT.0b013e3182076ff2.

Sioen, Isabelle A., Hse Pynaert, Christophe Matthys, Guy De Backer, John Van Camp, and Sterfaan De Henauw. "Dietary Intakes and Food Sources of Fatty Acids for Belgian Women, Focused on n-6 and n-3 Polyunsaturated Fatty Acids." *Lipids* 41, no. 5 (May 1, 2006): 415–22. doi:10.1007/s11745 -006-5115-5.

Smith, Earl L., Li-Fang Hung, and Juan Huang. "Protective Effects of High Ambient Lighting on the Development of Form-Deprivation Myopia in Rhesus Monkeys." *Investigative Ophthalmology and Visual Science* 53, no. 1 (January 2012): 421–28. doi:10.1167/iovs.11-8652.

Smith, Eric Alden. *Inujjuamiut Foraging Strategies: Evolutionary Ecology of Arctic Hunting Economy.* Transaction Publishers, 1991.

Smith, Terry L. *Celiac Disease.* Rosen Publishing Group, 2006.

"Solanine Poisoning." *British Medical Journal* 2, no. 6203 (December 8, 1979): 1458–59. doi:10.1136/bmj.2.6203.1458-a.

Solomon, Richard L. "The Opponent-Process Theory of Acquired Motivation: The Costs of Pleasure and the Benefits of Pain." *American Psychologist* 35, no. 8 (1980): 691.

Springer, Mark S., Robert W. Meredith, John Gatesy, Christopher A. Emerling, Jong Park, Daniel L. Rabosky, Tanja Stadler, Cynthia Steiner, Oliver A. Ryder, Jan E. Janečka, Colleen A. Fisher, and William J. Murphy. "Macroevolutionary Dynamics and Historical Biogeography of Primate Diversification Inferred from a Species Supermatrix." *PLoS ONE* 7, no. 11 (November 16, 2012): e49521. doi:10.1371/journal.pone.0049521.

Stanhope, J. M., and I. A. Prior. "The Tokelau Island Migrant Study: Prevalence and Incidence of Diabetes Mellitus." *New Zealand Medical Journal* 92, no. 673 (1980): 417–21.

"State-by-State Review of Raw Milk Laws." Accessed January 9, 2015. www .farmtoconsumer.org/raw_milk_map.htm.

Stec, James J., Halit Silbershatz, Geoffrey H. Tofler, Travis H. Matheney, Patrice Sutherland, Izabela Lipinska, Joseph M. Massaro, Peter F. W. Wilson, James E. Muller, and Ralph B. D'Agostino. "Association of Fibrinogen with Cardiovascular Risk Factors and Cardiovascular Disease in the Framingham Offspring Population." *Circulation* 102, no. 14 (October 3, 2000): 1634–38. doi:10.1161/01. CIR.102.14.1634.

Steele, Michael A. "Tannins and Partial Consumption of Acorns: Implications for Dispersal of Oaks by Seed Predators." *American Midland Naturalist* 130, no. 2 (October 1, 1993): 229–38.

Stevick, Richard A. *Growing Up Amish: The Teenage Years.* JHU Press, 2007.

Strachan, D. P. "Hay Fever, Hygiene, and Household Size." *BMJ* [formerly *British Medical Journal*] 299, no. 6710 (November 18, 1989): 1259–60.

Sugano, Michihiro, and Fumiko Hirahara. "Polyunsaturated Fatty Acids in the Food Chain in Japan." *American Journal of Clinical Nutrition* 71, no. 1 ( January 1, 2000): 189S–96S.

Sugiyama, Takemi, Ding Ding, and Neville Owen. "Commuting by Car: Weight Gain Among Physically Active Adults." *American Journal of Preventive Medicine* 44, no. 2 (2013): 169–73.

———. "*Trichuris suis* Therapy for Active Ulcerative Colitis: A Randomized Controlled Trial." *Gastroenterology* 128, no. 4 (April 2005): 825–32. doi:10.1053/j. gastro.2005.01.005.

Summers, R. W., D. E. Elliott, J. F. Urban Jr., R. Thompson, and J. V. Weinstock. "*Trichuris suis* Therapy in Crohn's Disease." *Gut* 54, no. 1 ( January 1, 2005): 87–90. doi:10.1136/gut.2004.041749.

Surbeck, Martin, and Gottfried Hohmann. "Primate Hunting by Bonobos at Lui-Kotale, Salonga National Park." *Current Biology* 18, no. 19 (October 14, 2008): R906–7. doi:10.1016/j.cub.2008.08.040.

Sutin, Angelina R., Roy G. Cutler, Simonetta Camandola, Manuela Uda, Neil H. Feldman, Francesco Cucca, Alan B. Zonderman, et al. "Impulsivity Is Associated with Uric Acid: Evidence from Humans and Mice." *Biological Psychiatry* 75, no. 1 ( January 1, 2014): 31–37. doi:10.1016/j.biopsych.2013.02.024.

Suzuki, Akiko. "The Okinawa Shock: As Life Expectancy Falls, World Watches with Bated Breath." *Asahi Shimbun: Asia and Japan Watch*. Accessed December 1, 2012. http://ajw.asahi.com/article/globe/feature/obesity/AJ201205270054.

Takasu, Nobuyuki, Hiroyuki Yogi, Masaki Takara, Moritake Higa, Tsuyoshi Kouki, Yuzuru Ohshiro, Goro Mimura, and Ichiro Komiya. "Influence of Motorization and Supermarket-Proliferation on the Prevalence of Type 2 Diabetes in the Inhabitants of a Small Town on Okinawa, Japan." *Internal Medicine* 46, no. 23 (2007): 1899–1904.

Tamakoshi, Koji, Hiroshi Yatsuya, and Akiko Tamakoshi. "Early Age at Menarche Associated with Increased All-Cause Mortality." *European Journal of Epidemiology* 26, no. 10 (October 2011): 771–78. doi:http://dx.doi.org/10.1007/s10654-011 -9623-0.

Tambalis, Konstantinos D., Demosthenes B. Panagiotakos, Stavros A. Kavouras, Sofia Papoutsakis, and Labros S. Sidossis. "Higher Prevalence of Obesity in Greek Children Living in Rural Areas Despite Increased Levels of Physical Activity." *Journal of Paediatrics and Child Health* 49, no. 9 (2013): 769–74.

Tan, S. H. *Saya Yang Tau*. New Straits Times Press (M), 1973.

Taubes, Gary. *Good Calories, Bad Calories: Fats, Carbs, and the Controversial Science of Diet and Health*. Random House Digital, 2008.

Tay, M. T., K. G. Au Eong, C. Y. Ng, and M. K. Lim. "Myopia and Educational Attainment in 421,116 Young Singaporean Males." *Annals of the Academy of Medicine, Singapore* 21, no. 6 (November 1992): 785.

Tehrani, Fahimeh Ramezani, Nazanin Moslehi, Golaleh Asghari, Roya Gholami, Parvin Mirmiran, and Fereidoun Azizi. "Intake of Dairy Products, Calcium, Magnesium, and Phosphorus in Childhood and Age at Menarche in the Tehran Lipid and Glucose Study." *PLoS ONE* 8, no. 2 (February 25, 2013): e57696. doi:10.1371/journal.pone.0057696.

Thien, Francis C. K., Jean-Michel Mencia-Huerta, and Tak H. Lee. "Dietary Fish Oil Effects on Seasonal Hay Fever and Asthma in Pollen-Sensitive Subjects." *American Journal of Respiratory and Critical Care Medicine* 147, no. 5 (May 1, 1993): 1138–43. doi:10.1164/ajrccm/147.5.1138.

Tilford, Gregory L. *Edible and Medicinal Plants of the West*. Mountain Press Publishing, 1997.

Todoriki, Hidemi, D. Craig Willcox, and Bradley J. Willcox. "The Effects of Post-War Dietary Change on Longevity and Health in Okinawa." *Okinawan Journal of American Studies* (2004): 52–61.

Trasande, L., J. Blustein, M. Liu, E. Corwin, L. M. Cox, and M. J. Blaser. "Infant Antibiotic Exposures and Early-Life Body Mass." *International Journal of Obesity (2005)* 37, no. 1 (January 2013): 16–23. doi:10.1038/ijo.2012.132.

Trepanowski, John F., and Richard J. Bloomer. "The Impact of Religious Fasting on Human Health." *Nutrition Journal* 9, no. 57 (2010). www.biomedcentral.com /content/pdf/1475-2891-9-57.pdf.

Trichopoulou, Antonia, Tina Costacou, Christina Bamia, and Dimitrios Trichopoulos. "Adherence to a Mediterranean Diet and Survival in a Greek Population." *New England Journal of Medicine* 348, no. 26 (2003): 2599–2608. doi:10.1056/NEJMoa025039.

Trinkel, Martina. "Prey Selection and Prey Preferences of Spotted Hyenas Crocuta Crocuta in the Etosha National Park, Namibia." *Ecological Research* 25, no. 2 (March 2010): 413–17. doi:http://dx.doi.org/10.1007/s11284-009-0669-3.

Turkington, Carol, and Deborah R. Mitchell. *The Encyclopedia of Poisons and Antidotes*. Infobase Publishing, 2009.

Turner, Christopher. "The Calorie Restriction Dieters." *Telegraph*, July 25, 2010. www.telegraph.co.uk/health/7898775/The-Calorie-Restriction-dieters.html.

"Turning the Food Pyramid on Its Head with Sally Fallon Morrell." *Off the Grid News*, episode 107. Accessed August 8, 2014. www.offthegridnews.com/2012 /06/21/turning-the-food-pyramid-on-its-head-transcribed/.

Tylleskär, T., H. Rosling, M. Banea, N. Bikangi, R. D. Cooke, and N. H. Poulter. "Cassava Cyanogens and Konzo, an Upper Motoneuron Disease Found in Africa." *Lancet* 339, no. 8787 (January 25, 1992): 208–11. doi:10.1016/0140 -6736(92)90006-O.

Uribarri, Jaime, Sandra Woodruff, Susan Goodman, Weijing Cai, Xue Chen, Renata Pyzik, Angie Yong, Gary E. Striker, and Helen Vlassara. "Advanced Glycation End Products in Foods and a Practical Guide to Their Reduction in the Diet." *Journal of the American Dietetic Association* 110, no. 6 (June 2010): 911– 16.e12. doi:10.1016/j.jada.2010.03.018.

"USA: Cultivation of GM Plants, 2013." Accessed August 11, 2014. www.gmo -compass.org/eng/agri_biotechnology/gmo_planting/506.usa_cultivation _gm_plants_2013.html.

USDA Economic Research Service. "Food Expenditures." Accessed August 24, 2014. www.ers.usda.gov/data-products/food-expenditures.aspx#26636.

Usui, K., T. Hiraki, J. Kawamoto, T. Kurihara, Y. Nogi, C. Kato, and F. Abe. "Eicosapentaenoic Acid Plays a Role in Stabilizing Dynamic Membrane Structure in the Deep-Sea Piezophile *Shewanella violacea*: A Study Employing High-Pressure Time-Resolved Fluorescence Anisotropy Measurement." *Biochimica et Biophysica Acta (BBA)-Biomembranes*, 2011. www.sciencedirect.com /science/article/pii/S0005273611003609.

Van Belle, Tom L., Conny Gysemans, and Chantal Mathieu. "Vitamin D in Autoimmune, Infectious and Allergic Diseases: A Vital Player?" *Best Practice and*

*Research: Clinical Endocrinology and Metabolism* 25, no. 4 (August 2011): 617–32. doi:10.1016/j.beem.2011.04.009.

Vardavas, Constantine Ilias. *Public Health Implications of the Mediterranean Diet: Its Interaction with Active and Passive Smoking.* Thesis, Maastricht University, 2010. www.researchgate.net/publication/49794345_Does_adherence_to_the _Mediterranean_diet_have_a_protective_effect_against_active_and_passive _smoking/file/32bfe50d2d7b16c5d0.pdf.

Vasconcelos, A. T., D. R. Twiddy, A. Westby, and P. J. A. Reilly. "Detoxification of Cassava During Gari Preparation." *International Journal of Food Science and Technology* 25, no. 2 (1990): 198–203. doi:10.1111/j.1365-2621.1990.tb01074.x.

Vassallo, Milo F., Aleena Banerji, Susan A. Rudders, Sunday Clark, Raymond J. Mullins, and Carlos A. Camargo Jr. "Season of Birth and Food Allergy in Children." *Annals of Allergy, Asthma and Immunology* 104, no. 4 (April 2010): 307–13. doi:10.1016/j.anai.2010.01.019.

Vassallo, Milo F., and Carlos A. Camargo Jr. "Potential Mechanisms for the Hypothesized Link Between Sunshine, Vitamin D, and Food Allergy in Children." *Journal of Allergy and Clinical Immunology* 126, no. 2 (August 2010): 217–22. doi:10.1016/j.jaci.2010.06.011.

Velasquez-Manoff, Moises. "What Really Causes Celiac Disease?" *New York Times*, February 23, 2013. www.nytimes.com/2013/02/24/opinion/sunday/what -really-causes-celiac-disease.html.

Vissers, Maud N., Peter L. Zock, Rianne Leenen, Annet JC Roodenburg, Karel PAM Van Putte, and Martijn B. Katan. "Effect of Consumption of Phenols from Olives and Extra Virgin Olive Oil on LDL Oxidizability in Healthy Humans." *Free Radical Research* 35, no. 5 (2001): 619–29.

Vitousek, Kelly M. "Caloric Restriction for Longevity, I: Paradigm, Protocols and Physiological Findings in Animal Research." *European Eating Disorders Review* 12, no. 5 (2004): 279–99.

Vizgirdas, Ray S., and Edna M. Rey-Vizgirdas. *Wild Plants of the Sierra Nevada.* University of Nevada Press, 2006.

Vlassara, H., and G. E. Striker. "The Role of Advanced Glycation End-Products in the Etiology of Insulin Resistance and Diabetes." *Touch Endocrinology*, 2010. www.touchendocrinology.com/articles/role-advanced-glycation-end-products -etiology-insulin-resistance-and-diabetes?page=0,2.

Vocks, E. "Climatotherapy in Atopic Eczema." In *Handbook of Atopic Eczema*, edited by Johannes Ring, Bernhard Przybilla, and Thomas Ruzicka, 507–23. Springer-Verlag Berlin Heidelberg, 2006. http://link.springer.com/chapter/10 .1007%2F3-540-29856-8_55#page-1

Vogel, Gretchen. "Genomes Reveal Start of Ebola Outbreak." *Science* 345, no. 6200 (2014): 989–90.

Von Mutius, E. "99th Dahlem Conference on Infection, Inflammation and Chronic Inflammatory Disorders: Farm Lifestyles and the Hygiene Hypothesis." *Clinical and Experimental Immunology* 160, no. 1 (2010): 130–35. doi:10.1111/j.1365-2249.2010.04138.x.

Vossen, Paul. "Olive Oil: History, Production, and Characteristics of the World's Classic Oils." *HortScience* 42, no. 5 (August 1, 2007): 1093–1100.

Wagh, Kshitij, Aatish Bhatia, Gabriela Alexe, Anupama Reddy, Vijay Ravikumar, Michael Seiler, Michael Boemo, Ming Yao, Lee Cronk, Asad Naqvi, Shridar Ganesan, Arnold J. Levine, and Gyan Bhanot. "Lactase Persistence and Lipid Pathway Selection in the Maasai." *PLoS ONE* 7, no. 9 (September 28, 2012): e44751. doi:10.1371/journal.pone.0044751.

Waite, Kathryn J. "Blackley and the Development of Hay Fever as a Disease of Civilization in the Nineteenth Century." *Medical History* 39, no. 2 (1995): 186–96.

Walker, Ronald, and John R. Lupien. "The Safety Evaluation of Monosodium Glutamate." *Journal of Nutrition* 130, no. 4 (April 1, 2000): 1049S–52S.

Wall, J. S., and K. J. Carpenter. "Variation in Availability of Niacin in Grain Products." *Food Technology*, October 1988, 198–204. http://ddr.nal.usda.gov/handle/10113/23799.

Walters, Dale. *Plant Defense: Warding off Attack by Pathogens, Herbivores and Parasitic Plants.* John Wiley & Sons, 2011.

Weick, Mary Theodora. "A History of Rickets in the United States." *American Journal of Clinical Nutrition* 20, no. 11 (1967): 1234–41.

Weinstock, Joel V., and David E. Elliott. "Translatability of Helminth Therapy in Inflammatory Bowel Diseases." *International Journal for Parasitology* 43, nos. 3–4 (2012): 245–51. doi:10.1016/j.ijpara.2012.10.016.

Westerdahl, Johan, Christian Ingvar, Anna Måsbäck, and Håkan Olsson. "Sunscreen Use and Malignant Melanoma." *International Journal of Cancer* 87, no. 1 (2000): 145–50. doi:10.1002/1097-0215(20000701)87:1<145::AID-IJC22>3.0.CO;2-3.

Westerterp, Klaas R., and John R. Speakman. "Physical Activity Energy Expenditure Has Not Declined since the 1980s and Matches Energy Expenditures of Wild Mammals." *International Journal of Obesity* 32, no. 8 (2008): 1256–63.

Weston A. Price Foundation. "Journal, Summer 2013: Our Broken Food System." Accessed August 8, 2014. www.westonaprice.org/journal/journal-summer-2013-our-broken-food-system/.

Weverling-Rijnsburger, Annelies WE, Gerard J. Blauw, A. Margot Lagaay, Dick L.

Knock, A. Meinders, and Rudi GJ Westendorp. "Total Cholesterol and Risk of Mortality in the Oldest Old." *The Lancet* 350, no. 9085 (1997): 1119–23.

"What Is CH-19 Sweet Pepper?" *Capsiate Natura*. Accessed October 9, 2014. www .capsiatenatura.com/whatisch_19sweetpepper.aspx.

Whitaker, Elizabeth D. "Bread and Work: Pellagra and Economic Transformation in Turn-of-the-Century Italy." *Anthropological Quarterly* 65, no. 2 (1992): 80–90.

Whittaker, John C., and Grant McCall. "Handaxe-Hurling Hominids: An Unlikely Story." *Current Anthropology* 42, no. 4 (2001): 566–72.

Whitten, Tony, Sengli J. Damanik, Jazanul Anwar, and Nazaruddin Hisyam. *The Ecology of Sumatra*. Tuttle Publishing, 2000.

Wiley, Andrea. *Re-Imagining Milk: Cultural and Biological Perspectives*. Routledge, 2010.

Willcox, Bradley J., D. Craig Willcox, and Makoto Suzuki. *The Okinawa Diet Plan: Get Leaner, Live Longer, and Never Feel Hungry*. Three Rivers Press, 2005.

Willcox, Bradley J., D. Craig Willcox, Hidemi Todoriki, Akira Fujiyoshi, Katsuhiko Yano, Qimei He, J. David Curb, and Makoto Suzuki. "Caloric Restriction, the Traditional Okinawan Diet, and Healthy Aging." *Annals of the New York Academy of Sciences* 1114, no. 1 (2007): 434–55. doi:10.1196/annals.1396.037.

Williamson, John A., Peter J. Fenner, John Williamson, Joseph W. Burnett, and Jacquie F. Rifkin, eds. *Venomous and Poisonous Marine Animals: A Medical and Biological Handbook*. University of New South Wales Press, 1996.

Wilson, Reid. "Maine Becomes Second State to Require GMO Labels." *Washington Post*, January 10, 2014. www.washingtonpost.com/blogs/govbeat/wp/2014 /01/10/maine-becomes-second-state-to-require-gmo-labels/.

Wjst, M., and E. Hyppönen. "Vitamin D Serum Levels and Allergic Rhinitis." *Allergy* 62, no. 9 (2007): 1085–86. doi:10.1111/j.1398-9995.2007.01437.x.

Wong, M.S., D. A. P. Bundy, and M. H. N. Golden. "The Rate of Ingestion of *Ascaris lumbricoides* and *Trichuris trichiura* Eggs in Soil and Its Relationship to Infection in Two Children's Homes in Jamaica." *Transactions of the Royal Society of Tropical Medicine and Hygiene* 85, no. 1 (January 1991): 89–91. doi:10.1016 /0035-9203(91)90172-U.

Woolgar, C. M. "Food and the Middle Ages." *Journal of Medieval History* 36, no. 1 (2010): 1–19. doi:10.1016/j.jmedhist.2009.12.001.

Worm, Boris, and Trevor A. Branch. "The Future of Fish." *Trends in Ecology and Evolution* 27, no. 11 (2012): 594–99.

Wrangham, Richard W. "Evolution of Coalitionary Killing." *American Journal of Physical Anthropology* 110, Supp. 29 (1999): 1–30.

Wyatt, Tom. *All Your Gardening Questions Answered*. Boolarong Press, 2012.

Yoshioka, Mayumi, Kiwon Lim, Shinobu Kikuzato, Akira Kiyonaga, Hiroaki

Tanaka, Munehiro Shindo, and Masashige Suzuki. "Effects of Red-Pepper Diet on the Energy Metabolism in Men." *Journal of Nutritional Science and Vitaminology* 41, no. 6 (1995): 647–56. doi:10.3177/jnsv.41.647.

Zabilka, Gladys. *Customs and Cultures of Okinawa*. 2nd rev. ed. Bridgeway Press, 1959.

Zhang, Jian. "Epidemiological Link Between Low Cholesterol and Suicidality: A Puzzle Never Finished." *Nutritional Neuroscience* 14, no. 6 (November 2011): 268–87. doi:10.1179/1476830511Y.0000000021.

Zhang, Y., L. Träskman-Bendz, S. Janelidze, P. Langenberg, A. Saleh, N. Constantine, O. Okusaga, C. Bay-Richter, L. Brundin, and T. T. Postolache. "*Toxoplasma gondii* Immunoglobulin G Antibodies and Nonfatal Suicidal Self-Directed Violence." *Journal of Clinical Psychiatry* 73, no. 8 (2012): 1069–76.

Zhernakova, Alexandra, Clara C. Elbers, Bart Ferwerda, Jihane Romanos, Gosia Trynka, Patrick C. Dubois, Carolien G. F. de Kovel, Lude Franke, Marije Oosting, Donatella Barisani, Maria Teresa Bardella, Finnish Celiac Disease Study Group, Katri Kaukinen, Kalle Kurppa, Markku Mäki, Leo A. B. Joosten, Paivi Saavalainen, David A. van Heel, Carlo Catassi, Mihai G. Netea, and Cisca Wijmenga. "Evolutionary and Functional Analysis of Celiac Risk Loci Reveals SH2B3 as a Protective Factor Against Bacterial Infection." *American Journal of Human Genetics* 86, no. 6 (June 11, 2010): 970–77. doi:10.1016/j.ajhg.2010.05.004.

Zhou, B. F., J. Stamler, B. Dennis, A. Moag-Stahlberg, N. Okuda, C. Robertson, L. Zhao, Q. Chan, and P. Elliott. "Nutrient Intakes of Middle-Aged Men and Women in China, Japan, United Kingdom, and United States in the Late 1990s: The INTERMAP Study." *Journal of Human Hypertension* 17, no. 9 (2003): 623–30. doi:10.1038/sj.jhh.1001605.

Zohary, Daniel, Maria Hopf, and Ehud Weiss. *Domestication of Plants in the Old World: The Origin and Spread of Domesticated Plants in Southwest Asia, Europe, and the Mediterranean Basin*. Oxford University Press, 2012.

# ACKNOWLEDGMENTS

I am indebted to the following people for offering me advice, guidance, and/or other means of support in the course of researching and completing this book:

**Australia:** Ashan Abeykoon and Greg Hampton (Charcoal Lane restaurant), Hung Ba Ha, Huong Dang, Jon Belling, Mark Olive, Nhan Nguyen, Su-Lyn, Thu Van Ha, Trang Nguyen, Van Dinh Luong, Vi Kinh Tran, Shanaka Fernando

**Belgium:** Matt Roosen (B&D Business Development)

**Canada:** Chi Tran, Chuck Brown and Michael Szemerda (Cooke Aquaculture), Dr. Dounia Daoud, Dr. Inka Milewski, Dr. Phuong-Anh Nguyen, Dr. Thierry Chopin, Efrain and Hee-Sun Andia, Gerry Oleynik (Kiefro Wild Boar Farm), Jim Gifford (HarperCollins Canada), Katie Schleit and Rob Johnson (Ecology Action Centre), Kyle Worsley, Michael Bandurchin, Ngoc-Tran Pham, Nguyen Ngoc Dung, Quoc Pham, Ross Horgan, Sieu Truong, Susana Hemken, Theresa 'Mai Khanh' Nguyen, Walter Henn (Bearbrook Game and Natural Meats), Wendy Burpee

**China:** Bengbu College, Grace, Jiangong Zhou, Kay, Micheal, Nancy Zhou, Shuyan Qi, Yifei, Ying Yuan

**Greece:** Agapi Fyssaki-Angelogiannaki, Costas and Ariadni Melengoglou, Nikolaos Kerdelas

**Iceland:** Ægir Freyr Stefánsson, Égzeus Belial, Elín Ýr Sigur-björnsdóttir, Kolbjörg Katla Hinriksdóttir, Linda Arnheiðardóttir

**India:** Bhanumathy Chandran, C.R. Chandran, Dr. Clea Chand-mal, Dr. Bajish Chandran, Dr. K. Jithendranath, Hyacinth Pinto, Subin Vazhayil

**Japan:** Albert Phu, Shinichi Motada, Yasuko Tsukamoto, Yoshida Yoko, Yoshiko and Hiromi Ono

**Kenya:** Heather Katcher, Huyen Ngoc Tran

**Laos:** Caroline Gaylord, Joy Ngeuamboupha

**Papua New Guinea:** Aloish, Dominic Anis, Frank

**Sweden:** Dr. Helena Pettersson, Dr. Linus Holm

**Thailand:** Amarat Chai, Amnat, Amy Phongphanh Chantha-vong, Dr. Jintana Yhoung-Aree, Hoang, Tasanee Athayu

**United Arab Emirate**s: Kristine Abante

**United Kingdom:** Anh Dao Nguyen, Anna Ohanjanyan, Dr. Jens Groth, Phan Nguyen

**United States:** Angélique Pivoine, Audrey Mai, Caitlin James-Le and John Le, Craig and LaVon Griffieon, Diane Ott Whealy, Dr. Cristina Moya, Dr. Dean Ornish, Dr. Gail Kennedy, Dr. Harish Rajagopalan, Dr. Jared Diamond, Dr. Matthew Gervais, Dr. Michelle Kline, Dr. Rebecca Frank, Dr. Stacy Rosenbaum, Dr. Tom Wis-dom, Emma Yee, Hee Sun Kim, Helene Atwan (Beacon Press), Jon and Soon-Young Hinkhouse, Jonathan Stutzman, Lindsay Parsons, Lisa Hao-Ran, Mark Sisson, Mary Ann deVries and Tom Schlife, Megan Pierce, Miho Morishita, Natalie Transu, Robert Faturechi, Ryan Langton, Sally Fallon Morell, Susan Rabiner, Thomas Greil-ing, Truong-Chinh Nguyen, Uyen Nguyen

**Vietnam:** Dr. Giap Nguyen, Dr. Huyen Thanh Vu, Dr. Luong Ly, Giang Huong Thi Vu, Hang Thang Le, Hang Thi Dao, Lap Dinh Nguyen (Dung Nhat), Ly Van Tran, Mac McDougall, Nhat Hoa, Phung Vong, Thanh Huyen Phan, Thuy (Trieu Son), Tran Thi Phuong Thao

I would like to give special thanks to Annie Nguyen-Bárány, Ina Melengoglou (www.altsys.gr), and Jana J. Monji for helping me with drafts of this work, and to Zhongxuan Li, Dr. Daniel M. T. Fessler, Dr. Guy Drouin, Dr. Stéphane Aris-Brosou; my father, Dr. Can D. Le; my literary agent Don Fehr (Trident Media Group), and my editor at Picador USA, Anna deVries, for their outstanding assistance.

# INDEX

# ABOUT THE AUTHOR

**STEPHEN LE** is currently a visiting professor in the Department of Biology at the University of Ottawa. He received a Ph.D. in biological anthropology from the University of California, Los Angeles, in 2010.